The Trouble with Science

The Trouble with Science

Robin Dunbar

HARVARD UNIVERSITY PRESS
Cambridge, Massachusetts

Printed in the United States of America

This Harvard University Press paperback edition published by arrangement with Faber and Faber Limited.

Library of Congress Cataloging-in-Publication Data

Dunbar, R. I. M. (Robin Ian MacDonald), 1947–
The trouble with science / Robin Dunbar.
p. cm.
Includes bibliographical references and index.
ISBN 0-674-91019-2 (pbk.)
1. Science—Philosophy. 2. Science—Social aspects.
I. Title.
Q175.D847 1996
500—dc20 96-17482

Contents

Preface

This book grew out of a course of lectures that I gave to undergraduates embarking on a degree in Anthropology. The book owes a great deal to the many people with whom I have discussed the ideas developed within it, in particular my former colleagues and students in the Department of Anthropology at University College London. In many ways the book owes a hidden debt to my own teachers: I should acknowledge a particular debt to John Crook (who taught me biology), Jeffrey Gray (who taught me psychology) and Geoffrey Warnock (who taught me philosophy). I am especially grateful to Celia Heyes, Julian Loose, Nick Maxwell, Henry Plotkin, Kim Richardson, Simon Strickland and Daisy Williamson for taking the time and trouble to read and comment on some or all of the chapters. Julian Loose deserves special thanks for his patience in shepherding the book through the editorial processes.

In the interests of making the book more accessible, I have not formally referenced sources within the text. However, I have added a bibliography which includes all the relevant sources which I hope will satisfy those with more stringent demands in this respect. I fear that many will feel that I have not done their disciplines full justice; in my defence, I can only point out that I have driven equally roughshod over my own research specialties. In many respects this will simply serve to emphasize the problems concerning the popularization of science that I discuss at some length in Chapter 8.

I
Introduction

Among a science teacher's most striking experiences are encounters with bright, eager students who are utterly unable to understand some seemingly simple scientific idea.

Alan Cromer: *Uncommon Sense* (1993)

In 1632 the Italian astronomer Galileo Galilei published his *Dialogue Concerning the Two Chief World Systems.* In doing so he inadvertently set in motion one of the greatest revolutions in the history of the human race. His achievement was to discredit, once and for all, the long-cherished view that the earth is the centre of a universe whose sole purpose is the sustaining of human life. The world, he told us, is not always as it seems. Overnight, we humans became bit-part players in a drama whose stage dwarfed us by its magnificence, in a plot for which we were at best a minor footnote. Galileo marked the end of a long haul up from the first glimmerings of a conscious thought in the mind of some prehistoric human ancestor a quarter of a million years ago to the triumphs of fully fledged modern science.

Although we have lived in the Age of Science ever since, we have remained ambivalent about Galileo's vision of the world. For the last 350 years we have continued to hanker after the cosy certainties of our intellectual infancy when we were the focus of attention, the purpose for which God had created the universe, the centre around which this enormous edifice revolved. With Galileo's book, we were thrust rudely backstage. Not surprisingly, perhaps, we have viewed the ever-rising tide of science with an ambivalence tinged with a growing sense of alienation, of no longer being in control of our destinies.

The trouble with science was born of these doubts, for Galileo's legacy spawned divided loyalties. On the one hand, the proponents of science, enthused by its dramatic successes, rushed headlong down the sometimes bewildering maze of corridors opened up by the scientific revolution. On the other, the reaction against the hard-edged world of science found expression in a yearning for a more emotionally sensitive relationship with nature. Many of those who nailed their colours to the Romantic movement's masthead in the nineteenth century, for example, did so in order to take a deliberate stand against the destruction of traditional values that science seemed to represent.

These concerns have not gone away. They underpinned the deep antipathy to science that prompted C. P. Snow's forthright essay *The Two Cultures* (science versus the arts) thirty-five years ago.

> The reasons for the existence of the two cultures are many, deep, and complex, some rooted in social histories, some in personal histories, and some in the inner dynamic of the different kinds of mental activity themselves . . . Western intellectuals have never tried, wanted, or been able to understand the industrial revolution, much less accept it. Intellectuals, in particular literary intellectuals, are natural Luddites.
>
> (Snow, p.22)

While the debate that followed the publication of Snow's Rede Lecture in 1959 clearly demonstrated that many within the humanities were highly supportive of science (and, indeed, endeavoured to apply the principles of science to the study of the arts), it did little to dispel Snow's point that a significant body of opinion existed within the intellectual community that was profoundly anti-science. In a curious way, this view was highlighted by Snow's observation that the word *intellectual* was, by common convention, never used to refer to scientists.

Snow, of course, overstated his case. And it would be no fairer now to insist that all who label themselves as either intellectuals or members of the humanities advocate anti-science views. Nevertheless, there is, I believe, growing evidence to suggest that this antipathy to science has, if anything, deepened as the humanities have perceived themselves to be increasingly beleaguered by the sciences. More disturbing still is the evidence that people, particularly those of school age, are being turned off science.

A Problem in the Making

One of the most alarming manifestations of this ambivalence towards science emerges from the statistics on science education. The National Science Foundation's Global Database on Human Resources for Science and Engineering reveals that, in 1991, a mere 4% of 22-year-olds in the United States held bachelors degrees in the natural sciences and engineering, compared with around 27% who held undergraduate degrees in all other fields (including social sciences). Of the 1,165,178 bachelors degrees awarded in U.S. institutions of higher education in 1993, a bare 15% were in natural sciences or engineering. Demand for science courses at universities generally lags far behind demand for humanities and arts courses, and the grade requirements for entry into science courses are invariably lower than those for humanities courses. The young are voting with their feet.

This situation is far from atypical. Throughout the developed world, the same pattern can be detected. In Europe, the proportion of all bachelors degrees awarded for social and natural sciences or engineering courses in 1990 was 50% in Britain, 30% in Sweden and 47% in France, in a year when the figure was 30% in the United States. Of the European countries, only Germany (at 72%) stands out as an exception to the general trend. Even the Pacific Rim economies have not escaped this fate: in Japan that year, only 40% of bachelors degrees were in these fields, while Taiwan managed 43% and South Korea a mere 37%.

The impact of these factors is particularly clear in the entry requirements for university courses. The situation in Britain is far from being unusual, so it can serve as an illustrative example. The average minimum qualification for conventional high school entrants into eight mainline science and engineering degree courses at eight leading British universities in 1994 was 18.6 points out of a maximum possible value of 30, with a range from 16 to 24 points. The equivalent for eight mainline humanities degree courses at the same universities was 22.8 points, with a range of 20 to 26. The lowest requirements were in chemistry, genetics and engineering, three subjects of fundamental industrial significance in the modern world. In effect, science students at these universities were nearly a whole grade class lower in ability right across the board.

Bear in mind that these entry requirements do not necessarily reflect the difficulty of the courses concerned: they are determined purely by market forces relating number and quality of applicants to the number of places available. Science courses have lower entry requirements because fewer high-quality students are applying for them.

It seems that many in the younger generation are being turned off the sciences at some point in their school careers, either because they find them too difficult or because they find them too boring. Either way, the risk is that the less able students will go for the sciences because it's easier to get a university place in these disciplines. The implications for the decades to come hardly bear thinking about. If the quality of science students is poor, then the quality of the science teachers in the next generation – as well as the quality of industrial research and development – will also be poor, for it is these same students who must fill these roles in the years ahead. Poor science teaching means a rapid downward spiral into ever poorer quality students in the decades to come. Poor-quality research means an inevitable downward slide in national industrial competitiveness.

We are, it seems, already witnessing the beginnings of that downward spiral. In his lecture to the 1995 meeting of the American Association for the Advancement of Science, Bruce Alberts, the president of the U.S. National Academy of Sciences, remembered how, as a graduate student in the early 1960s, he had been genuinely excited by the high school biology textbook he used while working as a volunteer teacher at a Massachusetts high school. Fifteen years later, his daughter brought home her first-year biology book from the San Francisco public school she was attending. Aware that this particular book was the best-selling school biology textbook in the United States, he opened it with a sense of anticipation.

"I could not believe what I was seeing," he said. "The book was basically composed of a list of biology words that had to be memorized by all students. There was absolutely nothing interesting about it: it was not teaching science, it was teaching science words."

The implications are not encouraging: teachers who do not understand science, and lack confidence in their ability to explain it, will never be able to inspire children in subjects that are intrinsically difficult to come to grips with. The prospect is a future generation that is scientifically illiterate.

Another gloss on the problem is provided by the surveys and polls on the public understanding of science. These polls consistently show that 70 to 80 per cent of the British public approve of science and think it does a good job for us. The consistency between the poll that was carried out by *New Scientist* magazine in 1984 and the two polls that were carried out by Gallup on behalf of the BBC in 1987 and the *Daily Telegraph* newspaper in 1992 is quite remarkable. Needless to say, the response to these findings was one of general satisfaction. What this self-congratulatory view overlooks, however, is the 15 to 20 per cent of the population who think that 'science attacks tradition and robs life of it spiritual meaning' (to quote one of the questions endorsed by 16 per cent of those polled in the 1992 *Daily Telegraph* survey). This apparently insignificant minority is, I believe, more important than the pollsters' glib assertions about the general public approval for science would have us believe.

The reason why we should be concerned about the minority that espouses anti-science sentiments is that many of these individuals are often better educated, more articulate and more committed to their particular views than the average citizen. A significant number of them are university-educated (though, of course, they tend to hold degrees in the humanities). More importantly, they frequently occupy influential positions within the social, educational and political establishments where they are able to wield a degree of political power that is out of all proportion to their number.

This contemporary dissatisfaction with science finds expression in two other phenomena characteristic of the late twentieth century. One is a dramatic resurgence of fundamentalist attitudes and beliefs, many of which are either self-consciously anti-science or actively seek to constrain its activities in radical ways; the other is marked by the emergence of philosophies of despair among intellectual élites within the humanities.

Almost all of the major religious faiths have experienced a startling growth in fundamentalism during the past two decades. Creation Science, with its vigorous anti-evolutionary stance, has been but one manifestation of this within the Christian tradition. Most other major religions have experienced equivalent developments. Beyond the more conventional religions, this same attitude finds expression in the kinds of New Age mysticism associated with revivals of, for instance, ancient Druidic ceremonial, witches' covens and the cult of Gaia (the mother-goddess said to represent the living earth), not to mention a flowering of millennial movements.

Barry Hugill, the *Observer* newspaper's education correspondent, was recently drawn to comment that 'in the United States, the coming millennium has spawned numerous cults, supported by thousands of apparently sane people. So nutty are their ideas that Noam Chomsky, the world-famous professor of linguistics, says he cannot rule out the possibility of a regression to "pre-Enlightenment times".' Although these cults are far less common in Britain, there is, he noted, 'evidence that many, especially the young, are flirting with ideas on the outer edges of reality'. Here is the next generation who will have the responsibility for lifting us out of the social and ecological morass into which we are inexorably driving ourselves; yet their beliefs about the world are being increasingly influenced by zany half-baked ideas whose relationship to reality is often at best tenuous.

Much of this has an understandable psychological origin. Indirectly, many of these movements have their origins in the same concerns that underpin the extraordinary success in recent years of both 'charismatic' movements within the conventional Christian churches and alternative medicine. The real world, with its diseases and its deep psychological stresses that threaten to overwhelm us, is often a genuinely frightening place. The staid structures of both the traditional churches and conventional medicine have done little to appeal to the emotional side of human nature, while science has inevitably seemed too intellectual and cold to fulfil that role.

It is no wonder that so many resort to faiths that promise to heal their ills or shelter them from the excesses of an uncaring world. There is a great deal at stake, for their commitment is often born of deep anxieties and fears. Understandably, perhaps, they frequently defend their chosen route to salvation with all the paraphernalia of fanaticism. *A propos* of alternative medicine, the writer Duncan Campbell commented in a recent article in the *Observer* newspaper: 'Who dares put down the chronically ill? Who dares question a panel of cancer patients? By this means, organized quackery has made the very idea of questioning its methods seem politically incorrect, and has thereby silenced its critics . . . Reporters who have looked objectively at "alternative medicine" have become the targets of hate campaigns and personal vilification.' Sadly, this defensiveness has made testing the claims of alternative medicine very difficult and this, in turn, has tended to polarize views on both sides.

The second form in which this phenomenon has manifested itself is in the rise of what, in the rarified atmosphere of intellectual circles in the humanities, has come to be known as 'Postmodernism'. Postmodernism, as its name implies, owes its existence largely to the collapse of the Modernist movement that dominated the arts and literature from the end of the nineteenth century. Dismay at the side-effects of modern technology, combined with despair at the horrors spawned by right-wing totalitarian regimes (many of which justified their activities on pseudo-scientific grounds) and the continued refusal of nations to co-operate with each other across the globe, served only to heighten the sense of disillusion that began to set in among intellectuals in the aftermath of World War II.

The resulting fragmentation of the largely forward-looking Modernist dream precipitated a pessimistic retreat into the view that there was no such thing as certain knowledge. Science itself came to be seen as one more expression of male-dominated Western cultural imperialism, a by-product of capitalism whose main function was to maintain the inequalities of the *status quo*. Within the humanities and the social sciences, such views have proved especially attractive to the more radical members of the younger generation, who have often used them as a powerful means of undermining the authority of their elders, many of whom hold more traditional (and conventionally scientific) views. In its most extreme forms within the social sciences and the humanities, the Postmodern influence often presents itself as the claim that there is no standard against which the validity of any idea can be tested. Its proponents cannot agree on a common agenda (because to do so would privilege some ideas above others), and the general thrust of the movement is self-consciously towards intellectual anarchy. In the social and historical sciences, it quite commonly takes the form of a wholly uncritical presentation of alternative views, irrespective of how rational or coherent these views might be. The reader is left to make of it what he or she will.

In contemplating this problem, it seems to me that all these different phenomena share a common element: an information gap of potentially disastrous proportions. Neither the proverbial man-in-the-street nor many of those who avow Postmodernist views in the humanities have any real understanding of what scientists do or how science works. Science has become a form of magic practised by an élite priesthood whose members have been subjected to a long and arduous apprenticeship in secret arts and rites from which the layman is firmly excluded.

Of itself, this need not be too serious: some of the more esoteric sciences might well be able to survive in quiet academic backwaters irrespective of whether or not the public understands them. However, their ability to do that does directly depend on the willingness of the public to finance science at all. A more serious problem is that, in our modern world, we have become totally dependent on the march of science to clothe and feed us. If we abandon science, we will, within decades if not years, be caught in Malthus's painful bind, for traditional forms of agriculture can no longer sustain the populations of the industrialized world. Because of their global scale, the resulting upheavals and social chaos will dwarf our past experiences of war, plague and famine.

What's Wrong With Science?

Let me spell out in a little more detail some of the misconceptions that have crept into public debates on science. One common perception is that all the man-made environmental disasters of the last century are the fault of science. In its more naive versions, the argument goes something like this. Science makes possible new technologies; therefore scientists are ultimately responsible for all the ills that result. So when the use of aerosol cans damages the ozone layer, thereby exposing us all to increased risks of skin cancer and global warming, or an ageing tanker runs aground and spills thousands of gallons of crude oil into the sea, it is science that is to blame because science produced the technology that made these things possible. When a nuclear power station leaks radioactive material into the environment, the responsibility lies not with those who mismanaged the technology or those who pushed machinery beyond its safety limits in the interests of marginally increased profits, but with those who discovered the physics that makes nuclear power generation possible. If scientists had never learned how to split the atom, or had been prevented from doing what they believed was possible, then we would not have had disasters like Five Mile Island and Chernobyl. Lest I seem to be setting up straw men here, let me simply add that I have heard this very claim being argued with deep conviction by both laymen and by academics in the social sciences. A very similar line is being vigorously peddled in some of the current media debates about artificial insemination of single women and post-menopausal motherhood.

An example of just this style of argument appeared in the London *Evening Standard* newspaper in the summer of 1991. A few days previously scientists in London had announced the discovery of the twenty or so genes on the Y chromosome that are responsible for maleness in mice; as a result, they had been able to create a 'transgenic' mouse that was genetically female but that exhibited all the normal male characteristics as a result of a bare couple of dozen genes that had been inserted into one of its chromosomes. This, argued the columnist Mary Kenny, would lead to the terrifying prospect of a drastically imbalanced sex ratio among humans because people would now choose to have all their female babies transformed into males. The spectre of Dr Mengele, the villain of the Nazi concentration camps, was brought into the story to imply a direct connection between modern genetics research and the activities of the infamous wartime medical researchers. The fact that people in all cultures throughout history have manipulated the sex ratios of their offspring (and by no means always in favour of sons!), and have not needed science to do it for them, went unnoticed. Nor, it seems, was this journalist aware that, under a well-established principle from Darwinian evolutionary theory known as Fisher's Theorem, the shortage of daughters would very quickly make them more valuable, and so encourage at least some parents to prefer daughters to sons, thus restoring the balance in the sex ratio within at most a generation or two.

To be sure, there are serious moral questions to be answered about our use and abuse of science, but we need to distinguish very carefully between questions about science and questions about the exploitation of scientific knowledge. After all, it is inconceivable that *any* human intellectual activity could be so totally free of the risks of misuse that it should be accorded a privileged position. If we accept strictures on the practice of science, we ought by the same token to accept similar restrictions on the arts, on the media, on religion, on politics and on all the other myriad facets of human culture. We would all, quite rightly, baulk at that. Yet precisely such action is often proposed with respect to science.

A rather different viewpoint is currently being promulgated in what is rapidly becoming a public debate conducted through the best-seller lists of major publishers. Brian Appleyard's book *Understanding the Present* is but one among many in this genre, but it will serve as a suitable representative.

The basis of Appleyard's argument is that science is socially and morally corrosive in the sense that it destroys the old certainties on which social life has depended for the past hundred millennia or so of human history. The sheer success of science – the overwhelming sense of power, of the ability to control the world, that it generates – has, he argues, destroyed our ancient dependence on the spiritual life. It is the emotional content of our spiritual side, he insists, that makes us human. With the advent of modern science, we have lost that sense of self, of humanity, that makes us what we are, that creates for us a purpose in life. 'Science . . . is spiritually corrosive, burning away ancient authorities and traditions. It cannot really co-exist with anything. Scientists inevitably take on the mantle of the wizards, sorcerers and witch-doctors. Their miracle cures are our spells, their experiments our rituals' (Appleyard, p.16).

Appleyard, it seems, has lost his sense of purpose in life and searches in vain for it among the bland accomplishments of science. All he can see is polished steel and glass, the sanitized paraphernalia of the modern science laboratory. It is a place in which he feels deeply uncomfortable, a stranger in a harsh and unfamiliar landscape. 'The problem is that science tells us there is nothing especially privileged about our position. Nothing is conclusive, we are eternally in "the middle of things" . . . There is nothing special about the way we happen to see things, nothing special about what the universe looks like from a human-sized perspective. In short, there is nothing special about us' (ibid., p.15). He surely speaks for many who are overawed by science's achievements, who are concerned about the apparent lack of control that they as individuals have over the world in which they live, who seek in vain for some meaning to their lives, who yearn for some deeper mystery in the universe to provide them with emotional comfort.

His answer is that we should abandon science as we know it because it dulls our appreciation of our own emotional inner life. Left to its own devices, so the argument seems to run, the cold logic of science will eventually eradicate music, art and literature, as well as those deep emotional stirrings we experience on viewing a scene of natural beauty or hearing a bird singing in the 'dawn chorus' in the half-light of an early summer morning. Instead, the world of science conjures up visions of the ordered society of Orwell's *1984* where 'Big Brother' tells you what to think, where life unfolds its drearily inevitable course in the dull grey world of an authoritarian state. Appleyard's clarion call is for a return to the inner world of the soul, where we can be in touch once more with our alienated feelings and emotions.

This is, of course, a view that harks back to the Romantic movement of the last century. It was also the view of many turn-of-the-century mystical philosophers. Here is the Russian P. D. Ouspensky, writing in the early 1920s:

> Developing science, i.e. objective knowledge, is encountering problems everywhere . . . There are multitudes of problems the solving of which science has not even attempted; problems in the presence of which the contemporary scientist, armed with all his science, is as helpless as a savage or a four-year-old child. Such are the problems of life and death, the problems of space and time, the mystery of consciousness . . . In the world lying beyond the domain of usual experience exact science with its methods has never penetrated and will never penetrate . . . We have also lost the understanding of magical ceremonies and rites of initiation into mysteries which had a single purpose: to help this transformation in the soul of man.
>
> (Ouspensky, pp.233, 235, 236, 257)

The undercurrents of conflict between science and anti-science are very real and the resolution of this conflict will have far-reaching consequences not just for the way we live, but also for our future political organizations, perhaps even for our survival as a species. Something, as the Bard remarked, is seriously wrong in the state of Denmark. We had better do something about it before it engulfs us.

In the chapters that follow, I try to tackle this problem head on. I explore both the nature of science and the reasons why, 350 years after Galileo, we still seem to have so much trouble coming to terms with it.

2

What is This Thing Called Science?

Experiments are the only means of knowledge at our disposal. The rest is poetry, imagination.

Max Planck

Scientists have seldom stopped to ask what it is that characterizes what they do. Being pragmatic people, they have simply got on and done it. Philosophers, on the other hand, have spent a great deal of time worrying about how we should define science and how we might distinguish it from religion (if indeed we can). Both groups have, in the end, been concerned with the same central issue, namely the certainty of our knowledge about the world, but their perspectives have been very different. Scientists have generally been more concerned with the validity of the particular inferences they draw about the world; philosophers have usually been more concerned with the nature of the scientific process as a whole. The proper place to start our enquiry, then, is with the philosophers.

The Art of Science

If modern science can be said to have had its beginning with Galileo in 1632, then the philosophy of science might be said to have had its beginning with the English philosopher and man of letters Francis Bacon. In a series of books published between 1606 and his death in 1626, he defended the cause of empirical* science and vigorously lambasted what he saw as the time-wasting triviality of the theologian-philo-

* Empirical is conventionally defined as relating to the observable physical world. In practice modern science deals with many phenomena which are not directly observable, including the fundamental particles of physics, genes, states of mind, etc. Indeed, even many phenomena in classical science were not directly observable (e.g. the force of gravity in Newton's physics). In this broader context, empirical refers to the use of data based on direct or indirect observation as the main way to find out about the world, and is contrasted with processes such as intuition, self-reflection or divine inspiration.

sophers of the medieval period. The issue at the heart of Bacon's onslaught against the medieval scholars was the certainty of knowledge. How can we be sure that our knowledge is completely reliable? The tradition established by the early Greek philosophers from Socrates onwards gave pre-eminence to logical deduction. For Bacon, this had been the root cause of all the unresolved (and unresolvable) squabbles that had dogged medieval philosophy for the best part of six centuries.

Bacon's contribution to the development of science lay in the fact that he identified the importance of both empirical observation and formal experiments as the only way of adequately testing hypotheses. His arguments proved to be extremely influential with the rapidly growing band of professional scientists over the next two centuries. In extolling the virtues of his new method, however, he was less than fair to his predecessors. To be sure, there had been plenty of hair-splitting to justify Bacon's outrage. But there were conspicuous exceptions, too, and Bacon owed a great deal to them. Among these, the names of Nicolas Oresme, Robert Grossetest, Duns Scotus, William of Occam (or Ockham) and Bacon's own namesake Roger Bacon were particularly important. They too had emphasized the importance of applying logic rigorously to arguments, had extolled the virtues of theories that demanded the fewest unproved assumptions (an important principle known to this day as 'Occam's Razor') and championed the empirical testing of hypotheses. As early as the first half of the thirteenth century, Grossetest was giving serious consideration to the problems of induction and the validity of knowledge. But even these giants of their day were outshone by the Arab alchemists and philosophers of the medieval period: men like al-Haytham, Muhammad ibn Musa al-Khwarizmi and Kamal al-Din al-Farisi raised the art of experimentation to new heights between the ninth and fifteenth centuries.

Bacon's vilification of Aristotle was even more misplaced. To be sure, Aristotle's works had suffered badly at the hands of the medieval scholars. Overawed by his formidable powers of analysis and his sheer breadth of knowledge, they had come to regard Aristotle as infallible. Unfortunately, he was known less through his own works than through the self-serving compilations made by the medieval scholars. It was not so much against the great polymath that Bacon railed as against the so-called medieval Aristotelians. For what Aristotle achieved in the fourth century BC was truly remarkable.

Though by no means the first empiricist among the Greek philosophers, Aristotle stood out among his contemporaries for the meticulous care with which he worked. He often complained that his predecessors' work was marred by careless observation. He also differed from them in one other fundamental respect, namely his insistence on the importance of using empirical observations to test hypotheses. Where his predecessors had tended to use observations merely as starting-points for their speculations, Aristotle brought his theories back to be tested against nature. Facts obtained by observation had pre-eminence above theories, he insisted. These two features were combined with one final ingredient that provided a powerful methodology: the use of rigorous logical deduction to develop causal hypotheses that could be tested with empirical evidence.

What lies at the heart of both Aristotle's and Bacon's approaches is the insistence on hypothesis-testing. But, in the centuries following Bacon, the 'scientific method' came to mean the 'experimental method', thanks largely to Bacon's vigorous campaign on behalf of the developing experimental sciences. Unfortunately, this is to confuse the general with the particular. Experiments are a particular way in which hypotheses can be tested, but they are not the only way. We can also test hypotheses by observation (as Aristotle did) as well as by assessing their internal logical consistency (as the great Greek geometers from Euclid to Pythagoras had done). Nonetheless, there is a difference between experimental and observational tests that Bacon well recognized, and this is the fact that observational data are fraught with many uncontrolled variables. Experiments have the crucial advantage that the scientist can control most of the variables except the one he or she happens to be particularly interested in.

In a world in which almost everything is influenced by many different factors, confounding variables are the bane of a scientist's life. Consider the simple problem of what determines the growth of crops. There is an almost infinite number of *possible* factors: the amount of rainfall, the temperature, the wind, soil type, slope and aspect of the land, the month when planting took place, the sign of the zodiac at planting, the number of birds on spring migration that year, the number of sacrifices made to the appropriate gods, the number of days since the world began – and that just lists some of the more obvious possibilities. There is nothing intrinsically wrong with *any* of these suggestions: they are all perfectly good scientific hypotheses. Our problem is to decide which are genuine factors that really do influence plant growth and which are incidental correlations completely unrelated to plant growth.

As it happens, we now know that the first five really do influence plant growth. The sixth, seventh and eighth are unrelated to plant growth as such, but correlate with variables that do influence it; and the last two are almost certainly irrelevant. But the point is that we are not to know this in advance. Were we to test only one hypothesis by checking our crops' performance, we might well get a good match between prediction and reality for all of these hypotheses. Using the sign of the zodiac as a guide to when we should plant crops, for example, might well have yielded very satisfactory results year after year in central Greece during the last few centuries BC. This is because, although the planets themselves have no influence on the growth of plants, their movements through the heavens do happen to correlate quite closely with some of the factors that do (notably the seasonal pattern of rainfall and temperature). But it would have proved to be a disastrous rule to follow in, say, South America at the time. Moreover, it would not help us today, even in central Greece, because the precession of the earth's axis of rotation (the gradual change in the direction of its alignment with respect to the polar stars due to the earth's wobble as it spins on its axis) has meant that the timing of the zodiac signs has moved round one whole sign since Aristotle's time: the sequence starts in the constellation Pisces rather than the constellation Aries as it did 2500 years ago.

Let me give a more specific example. Researchers at Israel's Tel Aviv University recently reported that people who sport thin moustaches are more susceptible to ulcers than anyone else. Now, this might tempt me to rush to the bathroom and shave off my moustache in the expectation that doing so will instantly reduce my susceptability to ulcers. Alas, such a course of action would be entirely useless: indeed, it might even have the opposite effect and *increase* my chances of developing an ulcer. The reason, of course, is that moustaches as such do not influence the development of ulcers. Rather, my personality influences both my susceptability to ulcers and my shaving habits. The clipped and neatly trimmed moustache is simply a correlated consequence of my general demeanour. Edgy people are more likely both to clip their moustaches and to develop ulcers. Leaving my moustache to grow long and unkempt or shaving it off altogether might in fact make *me* more edgy, and so more likely to develop ulcers. Correlations do not imply causes. The central problem in science, as in everyday life, is how to differentiate between real causal effects and the spurious ones that are due to confounding variables.

In Bacon's time, trying to unravel the complex web of relationships in real world phenomena was close to impossible without the aid of experiments in which only one factor was allowed to vary while all the others were held constant. For Bacon and those that followed him, observational data were the starting-point for theory-building, but that was all. Armed with an hypothesis based on observation, the scientist had to set about a series of rigorous experimental tests in order to rule out all the spurious correlations. However, the development of mathematical statistics over the past century or so has given us an array of powerful techniques that allow us to undertake equivalent tests on purely observational data. Statistical analysis, which uses mathematical techniques to separate out the influence of different factors, has made possible a dramatic rise in the number of non-experimental empirical studies, especially in the second half of the present century.

This change in emphasis has had important consequences in some areas of science, notably subjects like behavioural biology where experimental manipulations may easily destroy the very phenomenon that we seek to study. Until recently, for example, it was common practice to study the social relationships of monkeys and apes by convening groups of animals that were strangers to each other. However, the structure of many primate groups is extremely complex because it rests on sets of relationships between individual animals that have been built up over a very long period of time, sometimes extending back for several generations. Kinship, friendships based on long-term familiarity and frequent interaction, even a knowledge of relationships between third parties, have all been shown to be important contributory factors. Thus, although convening groups out of strangers can tell us a good deal about how monkeys build up new relationships, the lack of time depth (or history) means that some of the key components that structure primate groups are missing in these experimental situations. If those processes are important to the ways in which the animals build up their relationships, then their absence may radically alter the kinds of relationships that the animals develop. This in turn will influence the apparent structure of the group. From my own studies on gelada baboons, for example, we now know that when groups lack kinship networks because the animals are unrelated to each other, the monkeys prefer to associate with powerful high-ranking individuals, whereas in groups with well-developed kinship networks they prefer to associate with close relatives, irrespective of how these rank within the group.

Science, then, is a method for finding out about the world and not a particular body of theory. Whatever else science is used for, it is explanation that remains its central aim. Recognition of this has led some philosophers to draw a distinction between what the American philosopher George Gale has called 'cookbook science' and 'explanatory science'. The contrast recognizes that science consists of two distinct steps, namely the accumulation of empirical observations (packaged in the form of generalizations) and the invention of explanations that tell us why these generalizations exist. This is reminiscent of the distinction between 'knowing how' and 'knowing that' that the eminent philosopher Gilbert Ryle drew in his seminal book *The Concept of Mind*. Ryle argued that there is a crucial distinction between being able to say 'I know how to do *X*' and saying 'I know that *X* is the case [because . . .]'. The first implies technical competence, but only the second implies that the speaker knows why *X* works in the way it does. I would prefer to rewrite Ryle's aphorism in terms of 'knowing *that* [*X* is the case]' and 'knowing *why* [*X* is the case]', but the point remains the same.

The importance of this distinction can be seen if we return to our example of what makes plants grow. Using the signs of the zodiac or the arrival of the spring bird migration might well be a very effective rule for deciding when to plant one's crops. This is cookbook science: a set of rules that tell you what will happen, usually couched in the form 'If . . ., then . . .' ('IF you plant your crops when the birds arrive, THEN you will get a bumper crop in the summer'). But these are no more than rules of thumb: correlations based on generalizing many years' experience of events in the natural world. Such rules of thumb can be perfectly adequate for everyday purposes. The Egyptians, after all, used an astronomical calendar to predict the annual flooding of the Nile with a precision that we can hardly better today. For the Egyptians, such precision was essential because the duration of the flood was too short for the enormous army of labourers to be raised in time to make the most of the flood's agricultural benefits: they needed some way of predicting the flood long enough in advance to call up a widely dispersed workforce.

But neither the Egyptians nor our Greek farmer had any understanding of just why these events should in fact be correlated with each other (although they might well have had some theories of their own). If our farmer were to migrate to eastern Africa, disaster would ensue because the weather patterns in the tropics do not follow the same sequence they do in the eastern Mediterranean. Worse still, the birds do not migrate through the area in the same way they do in Greece. He would plant at the wrong time and be ruined. This was precisely the real-life story of the first European immigrants to the northern parts of the New World. The English settlers under Captain Smith who settled the Virginian coast in the seventeenth century suffered appalling conditions during their first few winters because they tried to implement farming practices developed in an entirely different eco-system in north-western Europe. Had it not been for the kindness of the native Indians (who bailed them out and taught them how to farm the local environment) and shiploads of stores and fresh immigrants from the mother country each summer, the colonial venture would have fizzled out in its first few years.

Only if we have a proper understanding of the mechanisms that drive the phenomena of nature can we predict the future with any certainty. And so it is that the ability to predict the future course of events, to control what happens (if necessary by experimental means), has come to be the touchstone of science. In a nutshell, explanation is the aim of science, and hypothesis-testing using empirical data is its central method.

Falsification, Revolutions and Programmes

In the centuries following Bacon, philosophers invariably concluded that the theories of science are simply generalizations derived from a series of observations. Having examined a number of examples of a particular phenomenon, I conclude that, for example, 'All swans are white' or 'Everytime lightning strikes, thunder always follows.' This gave rise to the idea that science has three separate stages: *description* followed by the *induction* of generalizations that then have to be *tested* against new observations of the same phenomenon (or perhaps an experiment) to check whether the generalization holds true. This view held sway among both scientists and philosophers right up until the end of the nineteenth century. Indeed, it dominated the social sciences as well as some areas of biology well into the present century where it went under the name of 'positivism' (the label given to this approach by the French philosopher–scientist Auguste Comte, one of the founding fathers of sociology). To most lay people even now, science consists in discovering new facts about the world. It was largely this view that provided the justification for the extraordinary flowering of biological and geological collections during the Victorian period and led, towards the end of the century, to the building of the great national museums in which to house these collections.

This essentially linear view of science was already being undermined even before it was formulated by the way some scientists worked. Gathering empirical generalizations might have been a good description of the way biologists and geologists worked even as late as the end of the nineteenth century, but the so-called Scientific Revolution of the seventeenth century had already provided a very different way of proceeding in the physical sciences. Here, generating explanations was the key to progress, not the proliferation of descriptive generalizations.

Meanwhile, the positivist view of science was not without its philosophical challenges. For, as early as the middle of the eighteenth century, the great Scottish philosopher David Hume had pointed out that the induction of generalizations faces a serious problem: the only guarantee we have for the success of the inductive method is its past success. But that itself is a generalization and, since the next example might disprove this particular generalization, we soon end up in a vicious circle in which we try to justify one generalization by another equally shaky generalization. Induction is thus fatally flawed and any form of empirical science based on it is necessarily weakened in consequence. Induction lacks the certainty of knowledge guaranteed by the deductive disciplines like logic and mathematics.

Undoubtedly the best-known attempt to solve this paradox is that of the Austrian philosopher Karl Popper. During the 1930s Popper had been particularly concerned to find some way of distinguishing between the statements of science and those of metaphysics (i.e. distinguishing statements that had some external validity from those of pure belief). Popper recognized that attempts to justify science in logical terms by reference to induction are inevitably doomed. Instead, he pointed out that scientists did not in fact simply accumulate instances of a given phenomenon and then derive generalizations from them. Rather, they generated hypotheses about the nature of the world (sometimes, but not always, from inductive generalizations) and they then submitted these hypotheses to rigorous testing. These tests, he insisted, were not attempts to *prove* a particular theory (a form of induction) but rather attempts to *disprove* it. Proof, he argued, is something that is logically unobtainable. We can only ever disprove something with any certainty for the very reasons that Hume pointed out: a single counter-example is enough to disprove a generalization, whereas proof would require the impossible task of documenting every instance of the phenomenon in question (including, presumably, those that have not yet happened!). Experiments, in other words, are designed to *falsify* the hypothesis under test, not to demonstrate its truth. And that, Popper insisted, broke the vicious circle of the problem of induction. So far from being the incubus of science, the occasional counter-example was precisely what the scientist was looking for: it was the very hallmark of good science.

This led Popper to coin the term 'falsification' to describe what scientists really did. Popper's conception of science as a process of falsification dominated the philosophy of science for the better part of half a century, and still remains influential among scientists. However, it eventually became clear that, in practice, scientists did not always follow Popper's principle. On some occasions, they seemed to accept hypotheses on little or no evidence; on others, they declined to reject hypotheses when the tests proved them wrong. In fact, Popper's principle turns out to be too stringent and would actually lead to the abandonment of science in a very short time if it were applied rigorously: scientists would soon run out of hypotheses to test simply because their knowledge of the world is too limited. This seemed to be a consequence of two shortcomings in Popper's account.

One difficulty for Popper's theory is the fact that much of science consists not in trying to prove theories wrong but in trying to define their limits of application by identifying the points at which the theories do not work (i.e. the areas in which they make incorrect predictions). Physicists are currently much exercised by the question as to whether the theories of modern quantum physics apply throughout the universe or whether they fail under certain circumstances (like the conditions that hold within black holes or during the catastrophic explosion of the Big Bang at the very origin of the universe 15 billion or so years ago). It is in order to try to replicate the peculiar conditions of the Big Bang that so much money has been spent building super-massive colliders like the CERN machine in Switzerland that is funded by the European countries: in these huge circular tunnels built deep underground, giant magnets accelerate subatomic particles to close to the speed of light and then smash them into each other. The resulting collisions release enormous quantities of energy, so allowing physicists to study the behaviour of particles under conditions that are more extreme even than those found in the fiery nuclear furnaces inside stars like the sun. If the predictions made by quantum theory about what should happen under these circumstances turn out to be wrong, then we will know that a new kind of physics will have to be developed to handle these extreme conditions. Quantum physics, like Newtonian physics before it, will prove to be only an approximation to the true underlying physics of the universe.

The second problem is that Popper's falsification procedure appears to be based on the view that causal relationships in the real world are simple 'one cause, one effect' processes. In reality, most phenomena in the real world are influenced (caused) by a number of variables, as we noted above in the example of the farmer and his crops. Were we to test the hypothesis that 'crops grow best if planted immediately after the spring rains' we might well find that they don't because we have not taken into account the type of soil where we did the experiment or the temperature at the time, both of which influence crop growth and may happen to counteract the influence of rainfall in the particular spot where we carried out our test. Popper had been influenced mainly by those disciplines like Newtonian physics where most phenomena have simple explanations based on a single key cause. When an apple is dislodged from its tree, gravity causes it to fall to the earth; day dawns as soon as the sun rises above the horizon.

Once again, it is the problem of confounding variables that stumps us in most real-life situations. Only in vary carefully controlled experimental conditions would Popper's Falsification Rule be a sensible one to follow. But even then it would work only providing we were omniscient and could identify all the confounding variables at the outset. And, of course, if we knew all that, there would no longer be any reason to bother with the experiment!

The beginnings of a solution to Popper's dilemma came during the 1950s. The American physicist-turned-historian-of-science Thomas Kuhn became interested in why physicists had refused to abandon Newtonian theory for so long during the nineteenth century, despite the accumulating evidence against it. From a detailed study of the history of physics, he concluded that science proceeds in fits and starts. Major new ideas eventually give rise to what he called 'scientific revolutions' when all the active members of a discipline suddenly agree on a new approach (or 'paradigm'). Once such a 'paradigm shift' has occurred, everyone settles down to a period of what Kuhn termed 'normal science' during which they probe and test the implications of the new paradigm. The aim during this period is to determine the new paradigm's 'boundary conditions' – the limits to its applicability. Eventually the predictions made by the new theory will begin to be falsified. At first, scientists will not immediately give up the theory. Rather, they will seek to defend it by invoking special *ad hoc* auxiliary hypotheses that explain why the theory should give different predictions in just those circumstances where it appears to make false predictions. But eventually the weight of falsified predictions would become so great that the theory would have to be abandoned. At this point, someone would suggest a new paradigm, a scientific revolution would occur and the whole cycle would start all over again.

Kuhn's conception of science appears to be in direct contradiction to Popper's and many people have viewed these two views as polar opposites. But in some ways this is to misunderstand the nature of their respective arguments. Popper's is a prescriptive statement of what scientists *ought* to do if they want to get things right; Kuhn's is a normative one about what they *actually* do in practice. Kuhn's description of how scientists work says nothing about whether a given theory or paradigm is correct or incorrect, merely that scientists tend to accept or reject it as a group. They might do this on the grounds that it explains the available evidence better than the old theory, or for some purely arbitrary reason (such as the outcome of a game of dice or for reasons of collective political belief).

It is not difficult to see how the last option might lead to the view that the theories of science are the product of the culture to which a scientist belongs and have no real external validity. Kuhn himself sometimes seems to want to adopt such a relativistic view. But on another interpretation, Kuhn's views are quite compatible with the rationalist view that scientists adopt new paradigms only once they have tested the old one to the point of destruction and have found a better one with which to replace it.

Precisely such an interpretation has been put forward by the Hungarian philosopher Imre Lakatos. He pointed out that scientists appear to behave according to Popper's view on some occasions, but according to Kuhn's on others. Lakatos argued that the apparent contradiction arises only because the philosophers of science had failed to recognize that these two cases involved radically different kinds of theories. Scientists, he suggested, worked in a multi-layered world in which some theories function in a programmatic way while other theories are more concerned with the details of how the programme itself works. A programmatic theory provides scientists with a reason for doing a particular experiment or with a particular way of looking at the world: it behaves like a Kuhnian paradigm. Within this programme, scientists generate subsidiary hypotheses that specify how the framework theory works in practice: it is these that scientists test in detail and accept or reject in a Popperian fashion. Darwin's theory of evolution by natural selection provides a framework theory for biologists: it encourages them to interpret their observations in a certain kind of way and suggests particular hypotheses to test. The subsidiary hypotheses may or may not be right, but their disproof is not itself evidence that the framework theory is wrong. It merely tells us that the framework theory does not produce its effects in quite the way we supposed. Let me give you a more specific example.

The theory of evolution provides us with a framework theory that allows us to make sense of the scattered fossil record. Until the early 1970s, it was widely believed that humans and chimpanzees last shared a common ancestor some 15 million years ago. This suggestion stemmed from similarities between the teeth of modern humans and those of an extinct group of fossil Asian apes called Ramapithecines. However, new techniques in molecular biology allowed Alan Wilson and his colleagues at the University of California to determine from comparisons of human and ape blood proteins that the last common ancestor probably lived as recently as 3–5 million years ago. A great deal of argument ensued, but in the end molecular biology won the day. The anatomists went back to look at their fossils again more closely and concluded that the Ramapithecines were in fact probably ancestral to the orang-utan, the only living Asian great ape. The mistake had come from relying too heavily on a single character, the thickness of tooth enamel, that Ramapithecines and humans happened to share because they occupied similar terrestrial environments. The tree of human evolution had to be redrawn, but the theory of evolution itself remained unaffected.

In fact, contrary to common popular belief, the theory of evolution cannot be disproved by any evidence from the fossil record: the fossil record can only tell us *how* evolution occurred and which particular pathways it took, not whether or not the theory of evolution is true. Disproof of the theory of evolution can only come through studies of the mechanisms of evolution (for example, natural selection), and these can only be done on living species. In trying to make sense of the fossil record, we assume that the theory of evolution is true, relying on other scientists to test the validity of the framework theory.

Lakatos also made an important practical point when he observed that there is no point in rejecting a framework theory just because there is evidence against it. Without a framework theory, we cannot ask questions or design experiments. So there is no point in abandoning a framework theory unless we have a better one to replace it with. Abandoning a framework theory in the absence of an alternative is about as useful as making a series of diary engagements when you don't have a calendar. It is much better to carry on using the old discredited theory until such time as an alternative appears. Indeed, the best way to find an alternative is to keep testing hypotheses generated by the old paradigm. By doing this, we at least have a chance of uncovering some crucial fact that will lead us to a new paradigm.

These changing perspectives on how scientists actually work led to an important reinterpretation of the relationship between theory and data. Recall that the eighteenth- and nineteenth-century philosophers interpreted the relationship as a linear one:

observations → hypotheses → tests

The scientist accumulates observations until he has enough to warrant drawing a generalization (an hypothesis), which he then tests against new observation.

The shift in views that occurred during the twentieth century led to a sharp split between the world of theory and the world of empirical data. As the great German philosopher Immanuel Kant noted more than a century earlier, theories are what allow us to make sense of what we see. An animal engaged in the act of mating has no intrinsic significance when seen in isolation: it is just one animal lying on the back of another. We can describe it well enough, but its behaviour only acquires significance once we have a theory of reproduction that tells us something about the *consequences* of that behaviour. An even clearer example of this is provided by particle physics: the scatter of lines that appear on a cloud chamber photograph in the aftermath of a collision between sub-atomic particles is, taken at face value, just a set of randomly drawn lines. They acquire significance (or, if you prefer, meaning) only when interpreted in terms of the theories of particle physics: one now becomes the track of a W^+ boson, another is a Z° particle, a third is a photon, and so on. Framework theories serve to direct our attention to particular phenomena in the observable world. They may be derived by induction from an accumulation of observations, but they need not be and, in most advanced sciences, they are not.

In effect, the new conception of science was circular rather than linear.

It involved two quite distinct but parallel worlds (the theoretical world in which theories reside, and the empirical world of observations) which are linked via a feedback process of hypothesis-testing:

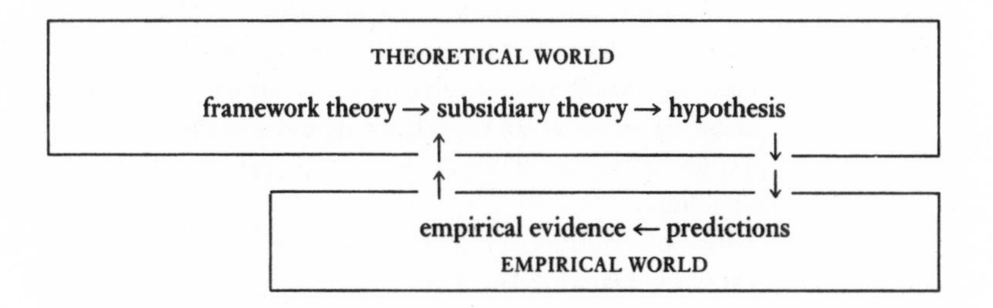

This conception of how science works is usually known as the 'hypothetico-deductive model', a rather ugly name given to it by the philosopher Carl Hempel. Theories are essentially constructs or models of how the world works. We work within a strictly theoretical world by deducing what consequences must follow from the model's assumptions and premises; we then test the validity of the model by comparing its predictions against the real world. So long as the model produces predictions that match what we actually see, we press on developing the model. But when the model fails to predict reality correctly, we alter the model accordingly or search for a better one. Science, in other words, is a feedback process: it learns from its own mistakes. Indeed, it behaves in a genuinely Darwinian fashion: only successful theories survive.

Of course, all disciplines involve induction in their early descriptive phases. But equally, any discipline that remains locked in this phase can do nothing except describe correlations in the world: it can never aspire to full scientific status by offering explanations as to why the world has to be the way it is. It remains a form of intellectual stamp collecting. Once a discipline moves beyond this descriptive phase, however, we find that theories exist in a theoretical rather than an empirical world: theories are developed from sets of assumptions that are, in principle, quite independent of the real world.

These two views of science were characterized as the 'bucket' and 'searchlight' models of knowledge, respectively, by Popper. Inductive (or cookbook) science is somewhat analogous to filling a bucket with spadefuls of sand (and hoping that something interesting turns up); formal (or explanatory) science works more like a searchlight by using the predictions generated by very specific theories as tools for exploring the real world.

If the searchlight model of science is true, then it greatly weakens the force of Hume's worries about induction. So far from being the centrepiece of science, empirical generalizations are at worst the starting-point for theory development, and at best the basis for testing a theory's predictions. This is not to say that theories in science are arbitrary. Our theories purport to describe the world as we experience it, and it is their ability to describe and predict the future states of that world correctly that is the hallmark of their validity.

The early history of evolutionary theory provides a nice example of these processes at work. Darwin's theory of evolution by natural selection constitutes an important framework theory in biology (in fact, it is the second most successful theory in science, after quantum physics). However, its operation in the real world depends on the existence of a mechanism of inheritance that enables it to have the effects Darwin inferred for it. Darwin had a great deal of trouble with this particular part of his theory and received much criticism in consequence. He changed his mind from one edition of the *Origin of Species* to the next and eventually settled for a form of inheritance which he called 'pangenesis'. According to this theory, each cell in the body contributes a tiny representation of itself to the sperm or egg and the contributions of the two parents are then literally mixed together at conception.

Darwin could hardly have been more wrong. Indeed, it was obvious to many of his contemporaries that his theory of inheritance was fatally flawed: it would quickly lead to the swamping of all the differences between individuals on which his theory of natural selection depended. In the end, half a century of intensive work in embryology and genetics undermined the pangenesis hypothesis completely. As a result, Darwin's whole theory of evolution waned in importance towards the end of the nineteenth century. However, the rediscovery in 1900 of Mendel's work on the inheritance of characters in plants revitalized interest in Darwin's theory: Mendel's theory of inheritance was precisely what Darwin's theory of evolution needed to make it work. Mendel had been able to show from detailed studies of pea plants that characters like seed colour and texture are inherited in particulate fashion: the seeds produced by crossing plants that have green and yellow seeds are either green or yellow, never a blend of the two colours. Moreover, green and yellow seeds always occur in the same proportions among the progeny of such cross-fertilizations. Mendel deduced that these characters are determined by factors (we would now call them genes) that are passed on at conception from parent to offspring, and that yellow and green represented different versions of the factor for seed colour. The factors are inherited intact from the parent plants and passed on intact to the offspring of the next generation.*

Thus Darwin's framework theory of evolution by natural selection remained intact (even if less widely discussed) despite the problems with the subsidiary theory for the mode of inheritance. A naive application of Popper's principle of falsification might have led to its complete abandonment in the 1870s. That it was not abandoned altogether was in part due to the implicit recognition (by at least some biologists) that it really consisted of two quite separate components: a theory about natural selection and the processes of evolution (which seemed to be basically right) and a theory about the mechanism of inheritance (which seemed to be wrong). What had caused much of the confusion was the fact that many of those interested in the mechanism of inheritance (including Mendel himself) had thought that they were studying the processes of evolution (i.e. an alternative to natural selection). Only once the logical structure of the theory as a whole had been straightened out did it become obvious what the solution was.

A Chaos Theory of Science

I have so far concentrated on the mainstream rationalist views of science. It would not be proper to end this brief overview of the philosophy of science without mentioning at least some other relativist views besides Kuhn's. In one important respect, the relativist view derives from Kant's assertion that our theories determine how we see the world. Taken to its logical conclusion, this argument insists that even our descriptions of the

* Mendel was lucky in his choice of pea plants: these happen to have very simple genetics. Had he tried the same experiments with humans, he would have had much more difficulty demonstrating the same effects. Although human skin colour is also determined in a straightforward Mendelian way by specific genes, there are many more such genes involved than in the case of pea plants. The results of adding together the effects of some eight different genes is to produce the colour shadings that are so conspicuous in humans. This makes it look like colours blending in the way Darwin supposed, but this is only a surface effect. The factors (or genes) that determine skin colour remain intact and can be passed on unchanged to the next generation.

world presuppose the existence of theories. In the social sciences this view underpins the claim that we are so imbued with language that the very words we use determine even the way we see the world around us.

It will be clear from what I have said already that there is a certain logical sense to this claim. If we interpret language in terms of our framework theories, then it is clear that language does indeed determine how we see the world: our framework theories are intended to serve precisely this function, namely to draw our attention to salient features in the observable world. Our language, so the argument would run, has evolved into the form it has because it provides us with a means of structuring the world in which we live. On this view, we all inhabit the same world, but use different words and ideas to talk about it: although this often leads to misunderstandings in cross-cultural conversations, the fact that we all inhabit the same world (and see it in much the same way) means that, once the linguistic differences have been clarified, we can hold a coherent conversation.

A more extreme interpretation is that different cultures literally inhabit different worlds because what we see and experience is actually determined by our language. The world of the Australian Aboriginal is necessarily very different from the world we Europeans inhabit because he uses different concepts and ideas to construct it; he has a different kind of logic and a different sense of causation and these make him *see* the world as being differently constituted to the way we experience it. Consequently there is no real hope that we can ever hold a coherent conversation with someone from another culture because we have no common basis (no concrete phenomenon 'out there') that we can use for translating from one culture's constructions to another's. This, in essence, is the version of the relativist stance that seems to have most heavily influenced many in social anthropology and the humanities.

One especially important relativist is the American philosopher Paul Feyerabend. Feyerabend's views are interesting because they are at odds with those of almost all the other leading philosophers of science. As a result, he has become something of a philosophical messiah for the sociologists of knowledge, while being regarded as something of a maverick by conventional philosophers of science. Feyerabend has always been very critical of philosophy of science as a discipline, going so far as to insist that it has been of no value at all to practising scientists because it has largely concerned itself with trivial problems of logic and meaning that do not impinge on the working lives of most scientists. That's a view that many scientists would undoubtedly warm to! Feyerabend's main importance in the present context, however, lies in the fact that he has challenged two crucial assumptions about the way we view science.

One concerns the way we choose our hypotheses. Feyerabend feels that science has become self-serving and that this tends to result in potentially interesting theories being rejected out of hand before they get a proper airing simply because they do not mesh with our current ideas about the world. Rather than deriving new hypotheses from our existing theories, he suggests we should consider any alternatives that occur to us, however outlandish they may seem at first sight. In this sense, Feyerabend advocates what he calls 'epistemological anarchy'. His other challenge to conventional theories about the nature of science is to argue that there is no such thing as a scientific *method*. Indeed, he even wants to claim that science as we practise it has all the hallmarks of religion: it has a standard set of beliefs that its advocates must adhere to or face ostracism and excommunication.

Few philosophers of science agree with Feyerabend's conception of science, though they acknowledge the force of some of his arguments. His claim that our knowledge of the world grows best when it faces direct challenges from new theories, for example, is in many ways uncontroversial. It is, after all, the basis of Popper's argument: Popper always insisted that conjectures (i.e. new hypotheses) should be as bold as possible on the grounds that the more implausible they are, the more powerful would be any resulting test. Feyerabend, however, wants to go further: his is a plea for intellectual pluralism, and it has inevitably proved to be particularly attractive to those who want to insist that science should not be specially privileged. Indeed, Feyerabend goes so far as to claim an equivalence between science and poetry: he once suggested that we should choose our hypotheses by the pleasure they give us.

As a general recipe for how to acquire knowledge about the world, however, Feyerabend's anarchic philosophy fails to take proper notice of the fact that scientists do try to use rational criteria to choose between competing hypotheses. Feyerabend would probably want to insist that when scientists decide to reject a particular theory, they do so largely on the basis of whim. And although this undoubtedly happens in some cases (after all, the sheer effort involved in thinking carefully through the assumptions, structure and implications of a complicated new idea are enough to make even an Einstein wilt), rejecting new ideas out of hand is not a particularly helpful way of doing science because progress in science ultimately depends on developing new ways of looking at the world. The real problem is that any fool can think up new ideas; the inconvenience of real life is that the key to progress lies in second-guessing how the world actually works, and that is an altogether much harder task.

The scientist's perennial problem is how to sift the handful of good ideas from the rest of the dross. Scientists in general operate a number of strategies that, in an informal way, help to do that. One is that the new idea has to convince a number of people besides yourself: that way, we at least guard against our frail human belief that every idea we dream up represents the last word in genius. Another is that the idea has to make sense in terms of what we already know: an outlandish idea is much less likely to be right if it contradicts our current knowledge about the world. Thirdly, the fact that a particular hypothesis is discarded because it does not make any sense does not imply that it should never be reconsidered again. One famous example of this concerns the Indian astonomer Subrahmanyan Chandrasekhar.

In 1928 Chandrasekhar developed some new ideas about stars' abilities to resist their own gravitational forces: he was able to show mathematically that stars would be unable to resist the gravitational forces of their own mass once they had exhausted their nuclear fuel, and so would collapse upon themselves. His calculations showed that for stars whose mass is less than about one and a half times the mass of our own sun (a value now known as the Chandrasekhar Limit), the result would be a super-dense star known as a 'white dwarf'. Chandrasekhar's calculations, however, also implied that more massive stars would collapse to vanishingly small size under the same circumstances (so giving rise to a phenomenon we now refer to as a 'black hole'). Unfortunately, his supervisor at Cambridge, the renowned physicist Arthur Eddington, was so shocked that he told Chandrasekhar the idea was absurd. Being a dutiful student, Chandrasekhar did as he was advised and forgot about his ideas. Several decades later, however, he resurrected these early ideas and worked them out in more detail. In due course, they earned him a Nobel Prize. Chandrasekhar's ideas have had an especially important influence on our current understanding of black holes and neutron stars. But, given the state of knowledge in 1928, they made no sense at all. However, it seems that a good theory cannot be kept down for ever.

Another example of this phenomenon is the theory of continental drift which was first proposed in 1915 by the German meteorologist Alfred

Wegener. Wegener argued that, 200 million years ago, the earth's continents formed a single large mass, the super-continent of Pangaea, which subsequently broke up into the continents as we now know them and have ever since continued to drift apart. His evidence for this was in part based on the fact that the continents as they are now do seem to fit together like a giant jigsaw puzzle. South America, for example, seems to fit rather neatly under the bump of West Africa's shoulder. Wegener's ideas were never taken very seriously until, during the 1960s, geologists began to discover evidence that the earth's crust actually consists of a set of plates. These plates float on the molten inner core of the planet, and are in constant motion. Where their edges collide, mountain chains like the Rockies, the Andes and the Himalayas are thrown up, or fault lines like the East African Rift Valley or California's San Andreas fault emerge. Once this had been appreciated, the biogeographical evidence for similar fossil species on continents that now boast very different faunas suddenly made sense. Wegener had been right after all.

We should beware of interpreting such examples as evidence that the scientific establishment operates a kind of Mafia-like conspiracy to prevent new ideas being heard. Most ideas that are of any consequence will survive or eventually be rediscovered. No matter how much like a Mafia boss an individual scientist may be, he or she cannot prevent everyone in the world from considering somebody else's tentative new theory. In fact, there are good grounds for seeing this critical process as an essential part of the scientific method by providing an initial filter to weed out those ideas that are either completely absurd or are still in need of a lot more careful development. New ideas are constantly being dreamed up by scientists and laymen. Thankfully, most never see the light of day, usually because they turn out on close examination to be logically flawed. Others are shelved pending more evidence or the solution of technical problems. Yet others are shelved because we cannot make sense of them with our current level of knowledge.

Science, then, is a process of intense criticism. Only a few ideas survive the first round of self-criticism on the part of the scientist who invents them and it is inevitably only these few that we ever get to hear about. However, it is all too easy to convince yourself that your latest brainchild is the cleverest idea since Einstein thought up the Theory of Relativity in 1905. The human species, alas, has never been short on self-appointed prophets with new or unusual ideas to peddle. The problem is how to avoid wasting too much time chasing every will-o'-the-whisp that comes along. The best way to choose between a worthwhile hypothesis and a less worthwhile one is to put it to the test in the arena of public debate. If your idea can survive the initially sceptical reception of your colleagues (each of whom would rather see his or her own pet idea as *the* theory of the day), then it is much more likely to be a promising one to pursue. If it is initially rejected, but you are convinced that it really is a good theory, then the likelihood is that you have not thought through the implications and assumptions clearly enough to convince the doubting Thomases. Rather than abandoning the idea, you should re-examine your assumptions more carefully in order to make a more convincing case for it. In the process, you will either discover its hidden flaws for yourself or rework your theory in such a way as to make the arguments in its favour more persuasive.

This dialetical process is an intrinsic part of science and it serves a useful function in preventing us from becoming distracted by too many ideas at once. And it is this that constitutes the most serious problem for Feyerabend's philosophy. Science generally progresses by the careful evaluation and testing of new ideas. Until that evaluation has been completed (and in the case of Newtonian physics it took the better part of two centuries!), there is little to be gained by chasing after every new idea that happens to come along. Without knowing the extent to which the old theory is wrong, we have no basis for deciding what a better theory has to be able to do. We simply would not be able to recognize it even when it was right in front of our noses.

So, although Feyerabend's radical suggestion may be a comfort for those who prefer armchair speculation to the hard work of exploring the real world, it is an unsatisfactory recipe for science if taken literally. And the reason for this has to do with our own inadequacies at the cognitive level: we find it incredibly difficult to think carefully through the maze of cause–effect relationships in complex theories and consequently are easily distracted by excitingly packaged but hopelessly flawed ideas.

The important point to emerge from all this is that science is a methodological prescription rather than a particular body of theory. It is a method for finding out about the world based on the generation of hypotheses and the testing of predictions derived from those hypotheses. Social anthropologists, among others, have argued that this approach is unique to modern Western culture. But is this really true? In the next two chapters, I shall try to show that the methods of empirical science are in fact genuine universals characteristic of all higher forms of life.

3
A Natural History of Science

Kaluli know to which trees the birds will flock in which season and which berry bushes to set traps around for marsupials.

Edward Schieffelin: *The Sorrow of the Lonely and the Naming of the Dancers* (1976)

Their errors are not wilful extravagances or the ravings of insanity, but simply hypotheses, justifiable at the time they were propounded, but which a fuller experience has proved inadequate.

Sir James Frazer: *The Golden Bough* (1890)

During the nineteenth century anthropologists like Frazer and Tylor had viewed the religions and myths of traditional societies as a kind of primitive pre-science. Myths, they argued, were attempts to understand and control the world. As attempts at explanation, they were undoubtedly rather primitive by the standards of modern science, but they were nonetheless bona fide attempts to explain the world as it was experienced.

This interpretation of myths continued to be advocated, albeit in modified form, during the 1930s and 1940s by anthropologists like Malinowski and Evans-Pritchard. Indeed, even as late as the 1970s, the influential French anthropologist Claude Lévi-Strauss was arguing an essentially similar claim.

However, from shortly after the turn of the century, a new strand of thinking began to gather momentum, which owed its inspiration to Emile Durkheim, one of the founding fathers of modern sociology. Durkheim argued that magic and religion have a rationale that is quite different from that of science. Beliefs and rituals, he insisted, are not intended to explain actions or emotions; they are not even statements about the physical world. Rather, they are simply commentaries on the nature of social life. Since magic and science cannot be mutually translated, they must require different kinds of explanation. It makes no sense, Durkheim insisted, to try to understand religion as an attempt to explain the unobservable, as the nineteenth-century positivists like August Comte had tried to do. If a ritual expresses a fear of God, it is because it is indirectly expressing a fear of the political power of the day (such as the very real local king).

During the 1960s, in particular, these ideas were taken up by a number of influential anthropologists, notably Edmund Leach, Mary Douglas and Oliver Tambiah. They argued that although the 'folk' natural histories of many non-Western societies appear to be descriptions of the natural world, they were in fact simply expressions of the people's own social concerns. Natural phenomena were attributed the characteristics of the particular society within which the people concerned were themselves embedded. Nature was thus used as a metaphor for the human condition or as an analogy or moral device in discussions of human behaviour. Whatever nature itself was or did, it was a closed book that human beings could not – or would not – read.

This claim allows for two alternative interpretations of science. One is that, while science obviously does work, it is not the case that all cultures practise 'science'. The world-views of 'pre-scientific' cultures, such as the Australian Aboriginals or the San Bushmen of southern Africa, are radically different from those of Western science: they have a different sense of causality and their view of the world has a logic all its own that is very different from the logic of science. Empirical science, in short, is something peculiar to the technologically advanced cultures of the modern world. The other interpretation is rather stronger. It suggests that modern science really is no different to the culturally constructed beliefs of these pre-scientific cultures: neither has any substantive validity as a description of the external world, which is, in any case, unknowable even in principle. Any successes that science might claim to have are purely accidental. This is the view most closely associated with the cultural relativists, those who argue, following Durkheim, that seen within their own light all cultures are equally valid and equally right: there is, and can be, no absolute truth.

It seems to me that the relativists' mistake lies in assuming that because some of the aboriginals' theories about the world differ radically from ours, it necessarily follows that all their theories differ from ours. To be sure, some of the origin myths favoured by traditional societies are quaint and implausible by the standards of modern evolutionary biology, just as our own biblical creation story seems quaint and implausible. But this does not mean that everything traditional peoples say about the natural world is equally wrong. In fact, much of their day-to-day biology is perceptive and, within the limits of their interests and technology, more than merely competent.

Is Science Pan-Cultural?

Let me begin by trying to convince you that genuine science can be arrived at independently of the intellectual traditions of the modern Western world. I shall do this simply by giving three examples from different historical and cultural environments.

My first example is the remarkable achievements of the Chinese during the first millennium BC. The Chinese never developed a fully fledged empirical science in the way we have it now, but they can claim some extraordinary technological firsts as a direct result of careful empirical observation of the world. Among these may be numbered methods for moving water around in order to irrigate fields, magnetic compasses for use at sea, gunpowder and the use of rocketry, printing and the use of ink, the invention of paper, silk production and water clocks, and a very detailed astronomy. The Chinese, after all, were making close observations of comets and other heavenly bodies long before we in northern Europe had even thought of the possibility that there might be more to life than ploughing fields and keeping the wolves away from our flocks. Indeed, in his book *The Crest of the Peacock*, George Joseph argues that, even as late as the 13–14th centuries AD, Chinese algebra was still the most advanced in the world.

We are so embedded in a culture that takes these inventions for granted that we find it difficult to appreciate just how remarkable these achievements really were in the first millennium BC. The preparation of paper from either wood or silk is no simple matter. Even the preparation of silk is not something that you or I could suddenly decide to do. It is a complicated process that depends on a very detailed understanding of the conditions required by the larvae of the silkworm moth for proper growth and pupation. That knowledge can only be acquired by close observation supplemented by trial-and-error learning.

Surprisingly, perhaps, the Chinese never developed a fully fledged scientific programme, and therefore their science remained strictly utilitarian (and hence 'cookbook') rather than explanatory. As has been pointed out by both Joseph Needham in his monumental history of Chinese science and by other commentators such as Toby Huff in *The*

Rise of Early Modern Science, the Chinese were impeded in this respect by a cultural tradition that prevented an exploration of the natural world for its own sake. Under Confucian influence (as the official State religion), their science was directed wholly to maintaining the natural harmony which the benevolent dictatorship of the empire was supposed to impose on its citizens. Science, in effect, became a tool of empire. Research was designed either to promote the economic base of the State's survival or to allow the State to predict the future (hence the interest in astronomy). Moreover, the development of science was totally subject to imperial dictat. When the imperial astronomers finally decided in 1447 AD that differences in latitude resulted in daylength in Peking differing from that in Nanking (the standard used to calibrate the water clocks in Peking), it required an imperial edict to allow daylength to differ in the two cities, and even then it was only issued because the matter was of some immediate importance to the imperial bureaucracy. Such constraints almost certainly prevented curiosity pursuing its nose.

My second example has to be Aristotle, the fourth-century BC Greek philosopher whose scientific achievements are probably unparalleled in the history of human thought. Aristotle was, above all, an empiricist. He insisted, in the face of deep cultural resistance from his fellow Greeks, on the primacy of hands-on research. Although his physics left a lot to be desired, his achievements in biology were formidable. The number of scientific 'firsts' to his credit reads almost like a dictionary of modern biology. He recognized that dolphins are mammals not fish (something not appreciated by European biologists until well into the nineteenth century); that the viviparous shark (a member of the dogfish family) gives birth to live young (not confirmed until the 1650s, and even then not widely known to European biologists until the 1840s); that reptiles can often regenerate parts of the body; that the yolk is in fact nourishment for the chick embryo (not the embryo itself, as everyone thought at the time); that the mammalian foetus is nourished via the umbilicus (and not, as everyone then believed, through sucking on the vili that line the uterine wall); that bees collect nectar from flowers; that hyenas are not hermaphrodites (a common belief right into the present century). He correctly described the nature and form of the Eustachian tube in the ear (not correctly described again until the Italian anatomist Eustachio himself did so in 1550). He got the essentials of bee sociology right (not formally described until 1740); he produced an account of chick embryology that was not bettered in European science until Harvey published his anatomical researches in the seventeenth century; he noted the hectocotylization of one arm of the octopus for use as an organ of insemination (still something of a biological puzzle); and he recognized that the sex of an embryo is determined at a very early stage in its development.

These achievements in purely descriptive science probably ought not to surprise us. Aristotle repeatedly stressed the importance of close observation, of dissecting organisms to see how they worked. If you look closely enough at nature, you just can't help seeing what is there. Even more impressive, perhaps, are Aristotle's achievements in what would now be classed as evolutionary ecology. He recognized, for example, that a species' birth rate is inversely correlated with its body size (though he got the explanation wrong), that larger animals need bigger ranging areas and that mortality rates are directly related to the effort put into reproduction. This is not to suggest that Aristotle had a theory of evolution (in fact, he believed in the fixity of species); but it is to suggest that he recognized the nature of adaptations (that species' characteristics are designed to enable them to survive in particular environments). Moreover, he made extensive use of the comparative method (later to become the foundation of Darwin's success) to derive generalizations, many of which have been rediscovered only in the second half of the present century. In addition, it has been claimed that his marine biology was so good that it was only superseded once marine biology became a discipline in its own right in the 1940s. More interestingly, perhaps, it has been suggested that Aristotle's success in this last respect derives from using much the same methods as those used by modern marine biologists (such as examining fishermen's catches repeatedly over a period of several years).

This paeon for Aristotle's science does not, however, mean that he got everything right. His physics, for example, was decidedly off-centre at times and he even made what we might regard as some startlingly crass mistakes in his biology. Thus, while he recognized that the embryo receives part of its biological input from the mother and part from the father, and he correctly identified the father's contribution as being contained in the sperm, he equated the mother's contribution with menstrual blood. He also thought that the lung was normally full of blood, but became drained on death to give rise to the pale thing we see on dissecting a cadaver. (Of course, it *is* full of blood, but since the blood is confined in the capillaries rather than the central space, we can't see it!) He also thought that small insects like mites and fleas generated spontaneously from mud (widely believed in Europe until proven otherwise by Francesco Redi in 1680); that the larva of larger insects like flies is the egg itself (in fact, the larva is the stage *after* the egg); and that eels don't breed but arise by spontaneous generation.

What is especially interesting about the things that Aristotle got right and those he got wrong is that they partition rather neatly into those things that were easy for him to see and those that were not. I combed Aristotle's biological works (the *De Partibus Animalium* ['Parts of Animals'], the *Historia Animalium* ['Natural History of Animals'] and the *De Generatione Animalium* ['Reproduction of Animals']) for specific facts that he got more or less right and those he got wrong. I then classified these in terms of whether or not he could easily have studied the phenomenon itself (depending on whether the crucial process can be seen with the naked eye or whether it takes place somewhere Aristotle simply could not have gone to). The results are summarized in Table 1. It does not take any sophisticated statistical analysis to see the correlation between these two variables.

TABLE 1 Aristotle's biological successes in relation to his ability to investigate them for himself

	Could Aristotle have studied them himself?*	
Number of topics that Aristotle:	YES	NO
got right	32	2
got wrong	2	10

* Defined as things that he could physically get to see for himself or that he could observe at first hand by dissection

Essentially, if Aristotle could see the thing and dissect it, he usually got it more or less right; but if he could not, he invariably got it wrong. One reason why he got it wrong in these cases is that he often resorted to the conventional wisdom of the day. As often as not, this was the product of idle speculation rather than careful observation. When he could see the chick embryo developing stage by stage by breaking open shells of different age, he could not fail to get it right. But he could not see fertilization itself taking place because the whole process takes place at the submicroscopic level; so he got key bits wrong. Nor could he have known that eels do in fact reproduce normally, because the European eel does this in the seaweed-tangles of the Sargasso Sea in mid-Atlantic (not to be discovered until Columbus inadvertently bumped into it on his way to the Americas in 1492). So while Aristotle's achievements are a tribute to empirical science, his failures are equally enlightening in that they remind us of the extent to which correct theories ultimately depend on good observations.

Aristotle was, of course, not the only ancient Greek philosopher to engage in empirical science. Yet, as Geoffrey Lloyd points out, the Greeks have rarely been credited with doing science. The problem, he suggests, is that much of the debate has focused on the question of whether or not they did experiments. That, as I showed in the last chapter, is a common misconception about science. However, even if it were reasonable to insist that science is only science if there are experiments, then, Lloyd insists, the Greeks still count. In the *Meteorologica*, for example, Aristotle describes an experiment to make rainbows by sprinkling water droplets through sunlight. Indeed, dissection (which Aristotle raised almost to an art form) can itself be considered a legitimate experimental technique since it involves interfering with nature. A few centuries later, the great astronomer Ptolemy discussed a number of principles relating to mirrors and reflection in his book on optics, and ended his discussion by listing a number of simple experiments that confirm these principles. Similarly, Hippocrates (usually considered to have been the founding father of modern medicine) described a somewhat gory experiment to determine whether or not liquid goes into the lungs when drinking. (It involved feeding stained water to a pig, and then slitting its throat to see whether the passages to the lungs were stained.) Later, by forcing water through the aorta, he showed that the heart's valves are designed to allow blood to pass in only one direction.

My final example is Islamic science during the ninth to twelfth centuries AD. At a time when European minds were trapped in the mental vice of early medieval Christianity, the burgeoning Islamic empires of the Middle East were carrying the torch of science that had been lit by the Greeks more than a millennium earlier. Not only did the Arabs preserve much of the learning of the classical world for us, but they also made major contributions in their own right to the modern scientific corpus. Active in centres of learning from Cordoba to Baghdad and beyond, they were responsible for significant advances in mathematics as well as in astronomy and chemistry. The eighth-century alchemist Abu Musa Jabir ibn Hayyam was probably the single most important figure in the history of chemistry. His penchant for trial-and-error experimentation (he was the first of the alchemists to recognize and emphasize its importance) led to the discovery of a number of new chemical reactions (including ones for the production of pure mercury and sulphur from naturally occurring mineralized forms). His theory that the differences between metals arose from the relative mixture of the two pure elements mercury and sulphur and their contamination with other impurities (itself a refinement of Aristotle's ideas) formed the basis of the phlogiston theory (the precursor of our modern atomic theory of chemistry) which was developed and refined in the seventeenth and eighteenth centuries by the German chemists Johann Becher and Georg Stahl, respectively. One of the most renowned intellectual figures of his day, Jabir's name was seldom absent from the many Arabic alchemical treatises written over the next four centuries. Through them, he found his way into the later European alchemical literature of the twelfth and thirteenth centuries, where his name appears in the Latinized form 'Geber'.

While the practical-minded Greeks gave us geometry, the more esoteric-minded Arabs gave us algebra. Indeed, the very word 'algebra' comes from the title of an influential treatise entitled *Hisab al-jebr w'al-muqabala* ('Calculation by Restoration and Reduction') written by the renowned Arab mathematician Abu Jafar Muhammad ibn Musa al-Khwarizmi that was published in Baghdad in 825 AD. Even the word 'algorithm' (now so central to computing) derives from the Latin translation of the title of one of his books, *Algorithmi de numero indorum* ['Calculation with Indian Numerals'], the original Arabic editions of which have long since been lost.

The impetus that the Arabs gave to this branch of mathematics proved to be absolutely vital for all later developments in science. Physics since the time of Newton would have been impossible without it; modern quantum physics, cosmology and chemistry, not to mention large sectors of nutritional biology (and its application to the agricultural sciences), evolutionary biology and demography would otherwise be quite impossible. Even the symbols we now use for counting derived from the Arabs, as their common name 'Arabic numerals' acknowledges. Their notation was a significant improvement on the cumbersome systems that had been in general use throughout Europe since the heyday of the Roman empire.

Among the other lasting contributions of the Arab mathematicians were systematic treatments of the basic rules of algebraic operations, solutions to non-linear simultaneous equations and the invention of decimal fractions. One of the leading Arab mathematicians, incidentally, was the algebraicist Omar Khayyam, better known in the West today for his lyrical poem the *Rubaiyat* but more famous among his contemporaries for his skill as a mathematician.

It is worth noting here that the Arabs were by no means the only ones who made significant contributions to modern mathematics. Much of their work was, in fact, made possible by developments originating in even more exotic cultures. They owed their numerical system, for example, to the Indian mathematicians of the closing centuries of the pre-Christian era. These ancient Hindu sages developed the decimal system of counting, invented the concept of zero and made a number of important contributions to both trigonometry and algebra (including methods for solving determinate equations and the beginnings of calculus – the latter some two millennia before it was officially discovered by Leibnitz and Newton in the West). The Arabs adopted these ideas, refined them and subsequently passed them on to us. The concept of zero, in particular, was of fundamental importance: modern science simply could not have been developed without it. Yet it was invented by Indian mathematicians two millennia ago.

On the more empirical side, we can point to several influential Arab scientists who made significant contributions to the later development of physics. The eleventh-century al-Haytham sought to introduce a new mathematical and experimental approach to the study of vision and light, and his book *Optics* was probably the most influential treatise on the subject until Newton's own book of the same title was published 700 years later. Al-Haytham's ideas were developed by Kamal al-Din al-Farisi (d. 1320), who used them in a series of experimental studies which demonstrated, among other things, that the rainbow consists of two refractions and a reflection of light within a water droplet. Al-Haytham's work was widely known to the early medieval theologian–scientists, including Bacon, Grosseteste and Theodoric of Freiburg (who independently replicated al-Farisi's rainbow experiment in about 1304). In fact, the later medieval Europeans often referred in their books to 'our Arab masters'.

In his book *The Rise of Early Modern Science*, Toby Huff emphasizes Copernicus's great debt to his Arab predecessors and argues, not without justification, that Copernicus's great breakthrough for modern science would have been impossible had he not had the benefit of developments made by the Arabs. He used the same graphical device, called a Tusi couple, to calculate planetary motions that had been developed by the astronomers of the Marâgha observatory in western Iran (established by the thirteenth-century Mongol ruler Huragu Khan). Moreover, his models for planetary and lunar orbits were based on those developed by the great Arab astronomer ibn al-Shatir (d. 1375) more than 150 years before. This is not to belittle Copernicus's great contributions to modern science (Copernicus had, after all, paved the way for Galileo through his revolutionary idea of inverting the Arab's geocentric model of the solar system to produce the modern heliocentric view); rather, it is to place him in his proper context as part of a great international community of scientists and thinkers whose uninhibited sharing of knowledge made it possible for science to advance rather than stagnate during the millennium-and-a-half that separated the heyday of ancient Greek science from the development of modern science that began with Galileo.

These three cases offer us examples where other cultures have made genuine contributions to modern science, or have come to the same conclusions that we have through careful observation. It is clearly important to recognize that these examples cover the full range of empirical science from pure cookbook science to genuine attempts at explanation. Much of Aristotle's biology, for example, is descriptive and many of the technological achievements of the Chinese were empirical rules of thumb. But other cases did involve attempts to explain why things were as they seemed to be. In many cases, the theories offered were plain wrong, but we should not denigrate them for that. That they turned out to be wrong was, often as not, simply a consequence of limited technology. In 1100 BC, for example, the celebrated Chinese mathematician Chou Kung estimated with surprising accuracy the obliquity of the ecliptic (the angle of the sun's apparent path through the heavens) relative to the equator, but his efforts to estimate the distance between the sun and earth were defeated by his failure to realize that the earth's shape was spherical rather than flat. The philosopher Ian Hacking points out in his book *Representing and Intervening* that progress in science has been impeded as often by the lack of technological know-how for making the right observations as by the lack of an adequate theory to understand what we observe.

Natural Science?

Convincing as these examples are, a sceptic might legitimately point out that they all derive from technologically advanced cultures. What evidence is there to suggest that scientifically less advanced cultures also engage in sound empirical science?

Here, for example, is a case of natural ethology at its best. In Japan there is a traditional method for fishing a small fish known as the *ayu* which is possible only because it exploits the fish's instinctive behaviour. During the breeding season, the males defend territories on the river bed, as do many similar fish all over the world. When an intruder enters a male's territory, the territorial male swims over and checks it out. If the intruder is female, he courts it; but if it is male, he delivers it a mighty blow with a side-swipe of his body. The Japanese developed a unique technique for fishing *ayu* known as *tomozuri* (literally 'decoy fishing') that exploits this behaviour.

The fisherman first catches a small male using more conventional methods. He then attaches this male to a large U-shaped hook such that the bottom half is buried in the bait-fish's body while the upper half runs up the outside. The baited fish is then played in the water using a very long light rod. By playing the decoy fish in such a way as to mimic the behaviour of an intruder, the fisherman is able to entice the territorial male to challenge, thereby impaling itself on the half-buried hook. This technique is used only for this particular species of fish. All other species are caught with conventional hooks and lines or with nets.

This technique is only possible because the fishermen who invented it understood the habits of this particular fish extremely well. That knowledge could only have come from detailed close observation: the fit between the fishing method and the fish's behaviour is simply too close to be a matter of chance coincidence. And it can only be used today by modern fishermen because they too have learned the habits of the *ayu* well enough to succeed in catching it. Naive fishermen invariably make an utter hash of an unnecessarily complicated technique.

My second example is likewise concerned with an intuitive appreciation of the behaviour of animals and how it can be exploited. The Fulani are a West African pastoralist people who have herded their long-horned cattle in the region of the upper Niger River for the past millennium. Dale Lott and his colleague Benjamin Hart were impressed by the way Nigerian Fulani herdsmen control their cattle by exploiting the animals' natural social behaviour. The herdsman inserts himself into the herd's natural social hierarchy by adopting the behaviour and privileges of a dominant animal. He uses unprovoked attacks on the more dominant members of the herd (especially the bulls), combined with more intimate grooming behaviour of the other animals, to establish and maintain his status as herd leader. In this way he can control the herd's movements during grazing as well as during travel between camp and feeding area. Lott and Hart point out that, unlike the horse-born stockmen of Europe and the Americas, the horseless Fulani cannot intimidate their cattle by physical size alone. Instead, they have learned to exploit the natural behavioural tendencies of the cattle that had evolved to facilitate the animals' own survival in groups long before domestication. If you doubt this explanation, I can only invite you to try taking the nearest herd of cattle from barn to field unaided.

My third example concerns the Boran, an East African pastoralist people. In common with many other tribes, the Boran exploit the behaviour of a small bird known as the honeyguide to locate the hives of wild honey-bees. The bees make their hives in cavities in rocky outcrops, abandoned termite mounds or hollow trees. The honeyguide locates these nests by following foraging bees back home. The bird's problem is that it cannot get into the nests to steal the honeycomb, partly because the honeycomb is usually secreted away in a space too narrow for the bird to enter and partly because it would be attacked by the bees if it tried. Instead, it searches for a herdsman out with his animals and attracts his attention by alighting on a nearby perch at head height and giving a characteristic call. Once it has succeeded in attracting the man's attention, it will guide him back to the nest. The man is then able to smoke out the bees and remove the honeycomb, but he always leaves a share for the honeyguide.

It is a symbiotic relationship in which both man and bird benefit from co-operation. But it only works because the man is able to understand the bird's behaviour and knows just what it signifies. He knows that the bird's fluttering flight around him and its peculiar calls mean that it has found a nest, not that it is trying to distract a predator about to stumble on its own nest (a form of behaviour that is common in many species of birds). He knows, too, that he has to move after the bird, so as to encourage it to show the way; he knows that he should wait and let the bird find him if he loses contact with it; and that he should reward the bird with some honeycomb if he wants it to repeat the process in the future. Furthermore, the Boran know that they can attract honeyguides to come to them by signalling their interest in finding honey: they do this by giving a characteristic high-pitched whistle that can be heard at distances of up to 1 km.

Intrigued by this phenomenon, the ornithologists Husein Isack and Hans-Ulli Reyer studied the relationship between Boran honey-gatherers and the birds in some detail. (In fact, Isack, a professional biologist by training, is himself a Boran and familiar with the honeyguide's behaviour from childhood.) The Boran claimed that they could tell not only the direction but also the distance to the bees' nest from the bird's behaviour. They claimed that the distance to the nest was indicated (i) by the length of time before the bird returns following its first disappearance after contacting the man; (ii) by the distance between successive perches where the bird waits for the following man; and (iii) by the height of the perches it uses (these get progressively lower as it approaches the nest). Isack and Reyer collected quantitative data on all three of these points, as well as a number of other aspects of the birds' behaviour (including the number of stops it makes and the total distance covered up to a given stop). They were able to confirm that only the three behaviours mentioned by the Boran themselves were significantly correlated with the distance to the nest at any given point along the route. Isack and Reyer were deeply impressed by the Boran's acute understanding of the behaviour of this particular species of bird.

My last example is a case of genuine experimental science in a traditional society. In this case, the people concerned are the Mende, an agricultural people from Sierra Leone in West Africa who cultivate the 'red' rice that is native to the region. Paul Richards has studied Mende agricultural practices in considerable detail. He found that the Mende engage in extensive experimental testing of new varieties of wild rice which are acquired from friends and relatives or by chance from wild crosses encountered in the forest or their own fields.

A Mende farmer will commonly reserve a small plot in a corner of his field to try out a new variety while planting his usual crop. In these cases, he will often measure out the rice grains with particular care and the new variety's productivity will then be compared very critically at harvest time with the performance of the rest of the field. Only if it is obviously better will he adopt the new variety. The Mende have a special word, *hugO*, for this kind of testing: it conveys a sense of testing that is quite different to that used in courts of law when a claim is put to the test (for which the litigious Mende use a completely different word). Richards notes that colonial agricultural officers recorded the existence of this practice in the 1930s and were of the opinion that it predated the arrival of colonial rule.

I contend that these examples all share a common feature: each one constitutes the end product of a process of empirical observation and informal hypothesis-testing. Indeed, the Mende example might even be considered a bona fide case of experimental science that is directly comparable to much of that found in the conventional agricultural sciences. The fact that we have not observed the processes of hypothesis-testing in most cases does not mean that they did not occur. It is difficult to see how the Boran could have achieved such a fine-tuned understanding of the honeyguide's behaviour without informally proposing hypotheses and keeping track of their successes and failures on a number of occasions. To be right on all five of the behavioural measures that Isack and Reyer looked at is statistically unlikely to occur by chance more than about once in thirty trials.

The Science of Survival

Ever since Marshall Sahlins published his influential book *Stone Age Economics* in 1960, social anthropologists have been inclined to view traditional societies as living in an ecological Garden of Eden. It is perfectly true that some societies (notably those on the Pacific islands) do seem to live under remarkably benign conditions. But even the inhabitants of idyllic coral islands cannot always have done so, for their arrival at newly discovered coral outcrops must have been attended by long-forgotten, but no less traumatic, periods of adjustment to the unfamiliar local conditions. That, at least, has been the experience of every migrant population within historical times, and it is difficult to see why it should have been any different for earlier migrants.

But there are other reasons for thinking that life on remote Pacific islands cannot always have been as blissful as it now seems. No less an authority than Malinowski himself recorded that many of these island societies were only too familiar with the prospect of famine and dreaded its periodic occurrence. On Hawaii the archaeological record reveals that the settlers who arrived around 500 AD had, by 1200, been responsible for creating a number of local environmental disasters by denuding the vegetation cover on the fragile inland habitats. By 1500, the populations on at least some of the islands (such as Kaha'olawe) had been forced to retreat to the coastal fringes or emigrate to other islands. Alas, it seems that the real world is not always a Garden of Eden in which Rousseau's innocent natives languidly while away the idle days between birth and death with minimal exertion and maximum hedonism. In the real world, survival places high demands on people's physical abilities as well as on their ingenuity and their knowledge of the environment.

The ability of the Australian Aboriginals to survive in an almost trackless desert is legendary. Recent studies by the anthropologist Neville White and the biochemist Kerin O'Dea have shown that populations pursuing a traditional way of life in the North Arnhemland region of Australia are extraordinarily healthy and surprisingly free from stress- and diet-related diseases, despite being much leaner than the official WHO guidelines recommend. What makes this especially surprising is that the Aboriginals make a culinary speciality out of fats: their favoured animal parts are just those that are known to be highest in cholesterol, namely the liver, the main fat depots and the brain. The Aboriginals show a fine-tuned appreciation of the habits of the animal species they hunt: they know exactly when during the year a species' natural fat-cycle is at its peak, and only then do they pursue that species.

They also show an appreciation of the general principles of nutrition, for they never eat meat without also eating carbohydrates (either in the form of honey or yams). 'Meat on its own turns the stomach sour,' they say. And, indeed, whatever else it does to you, a very high protein diet is anything but conducive to a long and happy life. As John Speth has pointed out, several of the early expeditions into the American hinterland starved to death despite a diet of unlimited meat culled from game-rich forests. Protein poisoning is a serious problem.

Biochemical analysis of the Australian Aboriginals' preferred foods demonstrates just how good their folk knowledge of the nutritional quality of local foods is. Their preferred species of yam (often extracted at great effort from deep underground) contains 30 per cent starch (compared to a mere 18 per cent in potatoes), as well as being unusually rich in fibre and essential trace elements like iron, copper and potassium.

O'Dea argues that their lifestyle is superbly adapted to the rigours of the environment that they have to cope with. Exercise is reduced to the absolute minimum, while at the same time they have an all-consuming hunger for fatty foods. In an environment that produces only very limited quantities of natural food, these predilections enable the Aboriginals to survive by minimizing energy expenditure while maximizing the intake of whatever natural energy sources they have available to them.

The North Arnhemland Aboriginals also practise a traditional form of natural farming (termed 'firestick farming') in which they deliberately set fire to whole tracts of bushland (often to the distress of European conservationists). The important fact from the Aboriginals' point of view is that, firstly, it flushes out a lot of the smaller game for them to hunt and, secondly, it encourages the growth of fresh vegetation as well as the germination of seeds (many of which have become adapted to regular firings and now require the intense heat of a fire in order to germinate at all).

The Aboriginals do, however, appreciate that it is easy to over-exploit resources in habitats as marginal as the Australian outback, and they normally replant a few seeds or tubers from the plants they harvest in order to ensure regeneration in the future. It would, nonetheless, be a mistake to conclude from this single example that all hunter–gatherers are natural conservationists, as some have tried to claim. There is ample evidence to show that the foraging strategies of at least some Amazonian hunter–gatherers have an adverse impact on the local availability of game and other resources. The issue here is really whether or not a given people recognizes the significance of conservation practices. They are likely to do so only once they start to suffer from the consequences of over-exploitation. This is likely to be much less intrusive in relatively rich environments like tropical Amazonian forest than in the kinds of marginal habitats occupied by Australian Aboriginals. This much is clear from the archaeological record for the Hawaiian islands: some populations were forced to change their subsistence practices from slash-and-burn horticulture to fishing because of the destruction of the habitat that resulted from their farming practices. Over-exploitation of local resources has been implicated in the downfall of more than one empire, from the city states of the Tigris basin in ancient Mesopotamia to the early Christian empire of Axum in northern Ethiopia.

A second example concerns the Maasai, a group of East African pastoralists whose wealth and survival depends entirely on the careful management of their large cattle herds. In order to survive in the unpredictable savannahs of eastern Africa, the Maasai not only have to understand the needs of their animals, they also have to be good ecologists. Biologist David Western has spent more than two decades studying the ecology of the Maasai in the Amboseli basin at the northern foot of Mt Kilimanjaro. This vast expanse of grassland is used not only by large herds of game animals, including elephants, but also by a substantial Maasai population who first invaded the area from the north during the early nineteenth century.

The cattle of the Maasai and other East African pastoralist groups often show striking variations in coat colour in different areas. Anthropologists have usually interpreted these as reflecting arbitrary preferences on the part of particular herdsmen. Virginia Finch and David Western found that the proportion of light- to dark-coloured animals varies consistently with altitude, with light-coloured cattle predominating at lower altitudes and darker ones at higher altitudes. From experimental studies, they were able to show that dark-coloured cattle suffer more temperature and radiation stress than light-coloured ones because they absorb more heat and more shortwave radiation. Dark cattle therefore have to drink more in the hot dry habitats at lower altitudes; consequently, the distance to which they can be taken for grazing is much shorter than it is for heat-tolerant light-coloured animals, and their productivity is much reduced as a result. Conversely, in the cooler habitats at higher altitudes, light-coloured animals die at higher rates during droughts, in part at least because they lose weight faster (and regain it more slowly) than dark cattle. In other words, there are clear biological advantages to the differences in coat colour that make one preferable to the other in a given habitat.

When Western later asked the Maasai about coat colour differences, it turned out that they knew perfectly well that dark-coloured animals have to drink more in hot dry habitats, and that light cattle are superior in hot habitats because they are more tolerant of heat stress. In addition, detailed questioning revealed that they had noticed that food intake is depressed when the animals become dehydrated, and had learned to use this as a guide to when they should water their cattle. They were also well aware that dark cattle grow faster than light ones under normal grazing conditions, a difference they attributed to the darker animals' greater food intake. They also thought that the higher mortality rates for light cattle at higher altitudes are due to the fact that the cold nights adversely affect both the physical condition and the milk production of the cattle. And, indeed, Finch and Western were able to confirm that dark-coated animals are not only able to conserve heat better, but they also warm up more quickly than light-coloured ones.

In another study, Western and Thomas Dunne looked at the Maasai's selection of settlement sites. They found that the Maasai paid very close attention to environmental characteristics and the effects that these have on the survival and productivity of their cattle and, ultimately, on their own families. The Maasai need to camp within easy reach of water both for their own use and for their cattle. However, the lower slopes of the ridges that rise away from lake and river beds are avoided because of the dangers of flooding. In addition, the upper slopes (especially on steep gradients) are also avoided because they exhaust already tired cattle returning to the safety of the camp at the end of a long day's trek. The preferred settlement sites on the middle slopes are thus a compromise between two complementary disadvantages. The Maasai also avoid light-coloured soils because these soils retain less heat during the night and are therefore less comfortable both for themselves and for the cattle (thermally stressed cattle produce much lower milk yields on these soils). At the same time, of course, settlement sites have to provide them with ready access to grazing for both their small and large stock animals, each of which has quite different requirements.

Western and Dunne commented that many of these features passed without comment among the Maasai themselves when the location of a new settlement was being chosen. The criteria are well known to everyone, as are the relative advantages and disadvantages of the various sites in a given area. Consequently there is no need to discuss them: they merely have to decide whether or not a particular site meets the balance of these criteria better than any other site. This often means that the only aspect to be discussed openly prior to a move is the social implications of the final shortlist of sites. The ecologically naive observer might easily be led to infer that only social and cultural factors influence Maasai settlement patterns, but these would actually be the last and least important matters to be considered.

On close questioning, it became apparent that the Maasai themselves are acutely aware of the environmental factors, and give virtually identical explanations to those deduced by Western and Dunne. Indeed, in several cases, it was the Maasai themselves that had pointed Western and Dunne in the direction of the correct ecological explanation. 'We found,' they noted, 'that the Maasai were most forthcoming on the subject of the environmental aspects of site selection when we accompanied them to the areas concerned and when they became aware of our ability to discuss the environment in a manner that related to livestock and human welfare' (Western and Dunne, p. 95).

Without a scrap of formal education, the Maasai had got the correlations right; better still, they had even managed to deduce the right explanations for many of these correlations. But why should this surprise us? The Maasai have spent many centuries living in these environments. They have enormous experience of how their cattle perform under different conditions. A man's cattle are of overriding importance to him, both for his family's survival (as the Maasai are only too well aware) and for his own social status within the community. When things are that vital, you take good care to ensure that they do as well as they possibly can. And that necessarily involves paying close attention to how they respond to changing environmental conditions.

Studies of traditional societies have commonly revealed that their knowledge of the natural world is in fact remarkably good. Jared Diamond, for example, found that the way one New Guinea tribe classifies birds is virtually identical to the way ornithologists do. Similarly, Aubut and Newsky found that the Bambara people of Mali in West Africa have a system of soil classification that is very similar to that developed by Western geologists to describe soil textures. Indeed, Scott Atran has argued in his book *Cognitive Foundations of Natural History* that natural history classifications are in general all remarkably similar (and are based on similar principles and criteria of classification) across all human cultures, despite the enormous cultural differences between the peoples concerned. He argued that humans have a natural tendency to classify the living world into types (i.e. species) that are based on their natural properties. The radical differences between the various cultures makes a common cultural heritage an unlikely explanation for the similarities in their classifications. Consequently we are left with no alternative but to assume that these similarities arise from the fact that the different peoples are describing the same world using the same basic principles.

The Pokot and Turkana pastoralists of northern Kenya know about the seasonal availability of various plants and recognize their different roles in stimulating milk and meat production in their animals. Like many other pastoralists, the Wodaabe of the Niger region know that certain plants are toxic to their animals; they know, for example, that the plant *Zornia glochidiata* is only toxic in its early vegetative stages, but quite harmless otherwise. They know that some plants, such as the shrub *Ipomoea acanthocarpa*, are salty and deliberately take their cattle to them during the rainy season for the same reason that European farmers provide their cattle with saltcakes. (It enhances the animals' ability to extract nutrients from their fodder.) The Rendille of Kenya differentiate clearly between the needs and preferences of their different livestock, recognizing that their camels prefer silty chalky soils and their cattle lateritic or montmorillonitic soils, whereas their sheep and goats prefer hard stony ground.

The Fulani pastoralists of Mauritania have a well-developed system of water lore. Like Western hydrologists, they use a variety of indicators for locating underground water: these include the local topography (shallow acquifers are usually found near natural ponds or in depressions in mountains), plant species (especially the presence of tap-rooted trees like *Bauhinia*, *Tamaris* and *Acacia albida*), the vigour of the plants (especially the greenness of their leaves in the dry season) and the presence of certain animals (such as wild boar which can only live where they can dig for roots, or amphibious lizards and tortoises that depend on water). They are also familiar with geological strata: they know that, in order to find water, they must dig down through the layers of red or grey clays until they find a sandy layer.

In his detailed study of the agro-pastoralist Shona of southern Zimbabwe, Ken Wilson found that these people use a large number of traditional signs in deciding when and where to plant their crops. They pay special attention to the form and direction of the prevailing winds (arguing correctly that winds coming with the northward drift of the Inter-Tropical Convergence Zone provide the most persistent rainfall) and to the timing of the onset of the rains (in Zimbabwe, at least, the rains are shorter if they start after mid-November). They also say that very high temperatures are necessary for good rainfall (a point generally accepted by meteorologists). Other claims made by the Shona have not yet been tested and may or may not prove to be correct. These include their claim that the position of the Pleiades constellation indicates whether the following year will be one of famine or plenty, and the claim that certain trees (notably *Lannea*, *Diospyros*, *Mimusops* and *Uapaca*) fruit more heavily in drought years.

Wilson also found that the Shona exploit the different soil types with a fine attention to their natural characteristics. They switch both seasonally and annually between sandveld and clayveld zones according to circumstances so as to obtain the best growing conditions for their crops. They know that sandvelds suffer from a nutrient constraint, so that they require manuring if their use is to be effective, whereas clayvelds have a water constraint so that large areas have to be ploughed and planted quickly to make best use of them when the rains come.

Not surprisingly, perhaps, traditional peoples' knowledge of their environment is often closely related to their needs. The Bambara's soil classification, for example, is not equally detailed for all soil types, but varies according to the importance of the soil type for their economy: it is most detailed for sandy soils because these are used for cultivation. Similarly, the !Kung San bushmen of southern Africa have a very detailed knowledge of the behaviour of the animals who share their environment, but their level of knowledge is very much dictated by what is commonsensical or important for their survival. They do not know, for example, whether the lion cub's eyes are open or closed at birth (they say it is foolish to approach a lioness too closely when she has cubs merely to find out); but they do know that migrating quelea weaverbirds will strip the leaves off bushes and trees a few days before settling in their thousands (a cue that Western agriculturalists learned to use from the !Kung). Biologist Nick Blurton-Jones has commented that the !Kung have a fine-tuned sense of the reliability of knowledge: they are reluctant to make inferences on the basis of hearsay, preferring direct observation whenever possible.

One last example that has long intrigued Westerners concerns the formidable powers of navigation displayed by those peoples that undertake long-distance migrations. Among these, the navigational skills of the ocean-going Micronesian islanders of the western Pacific are legendary. Using traditional lore handed down from navigators to their apprentices over many generations, they can sail courses that are virtually compass-perfect on a 4,500-mile journey during which they are out of sight of land for more than thirty days. While it is clear that the navigators rely on piloting between known landmarks during the early and late stages of a voyage (and often use knowledge of distinctive wave formations created by underwater structures when out at sea), much of the long-distance navigation between major landfalls is done on the basis of celestial navigation using traditional knowledge of the star patterns above individual islands. They know, for example, that the star we call Altair is directly above an observer on the island of Puluwat, while Sirius lies above Fiji and Arcturus above Hawaii. Knowing where these stars lie when they reach their zenith allows you to work out the direction in which to head in order to reach any given island.

This is no mysterious skill, simply the product of hard-earned knowledge accumulated across the centuries and handed on from one generation to the next. If they seem to involve feats of memory that we find astonishing, that is merely a reflection of a laziness on our part that derives from the luxury of being able to record everything we need to know in books or on computer disks. In real life the ability to learn large quantities of information quickly can mean the difference between life and death. Many years ago I attended a lecture by a trans-Saharan expedition that provided a graphic illustration of this. Though confident of their ability to find a way across the Great Sand Sea in the Libyan Desert, they nonetheless hired an elderly Arab guide in order to avoid wasting too much time trying to find passable routes around major obstacles. The guide claimed to have crossed the Sand Sea once before as a young man. Giving him full reign to direct the leading vehicle, they checked his route against their own instruments as they went along. Using only environmental cues like the colours and shapes of sandy areas and rocky outcrops, he took them on a dead straight compass bearing across a trackless waste. Even if he had done it more than once, it was a formidable *tour de force*, because a marginal error on his part would have left them hopelessly lost 1,000 miles from either water or safety. When your very survival depends on learning the relevant features of the environment, you get to learn them fast. Those who do not seldom get a second chance.

Despite frequent claims by contemporary anthropologists that traditional peoples do not engage in science, it is hard to interpret any of the examples given in this chapter in any other way. Only if we make the mistake of assuming that science is a particular body of theory can we argue that these peoples are not engaged in science. But, as I argued in the previous chapter, science is not a body of theory: it is a method for finding out about the world that combines empirical observation with causal inference. Indeed, if science really is a particular body of theory, then we would be forced to conclude that when Newton was developing his theory of gravity in the late seventeenth century he was doing science, but, seen from the perspective of the late twentieth century, he was not, because quantum and relativity physics have shown Newton to have been fundamentally wrong. His theory turned out to be a crude approximation – good enough to get by in everyday life, but not good enough for predicting the outcome of very precisely controlled experiments. This is clearly absurd, if only because Newton is universally acknowledged to be the paradigm example of a scientist. The mistake, of course, comes from confusing a method with the particular results that it yields at a specific point in time.

Within the technological constraints of their economies, all the peoples I have discussed are clearly engaged in very sound empirical science. To be sure, most of it is cookbook science, based on the recognition of regularities in the phenomenal world. But that in itself is the foundation of all modern science. To castigate traditional cultures for failing to go beyond cookbook science to develop full-blown explanatory science is absurd: they do not need explanatory science as we know it to get by in the world. Indeed, it would be counter-productive for them to waste the time and effort needed to go beyond cookbook science and would very likely place their survival in the world at risk. For them as much as for us in the West, it is the correlations of cookbook science that are important for everyday life. Even we don't need to know about quantum physics to get by in the world, so why should we expect traditional peoples to need to delve into the minutiae of these phenomena?

Nonetheless, the anthropologist Claude Lévi-Strauss has argued that pre-scientific societies have not always stopped short at cookbook science, and have often made attempts at explanatory science as well. Few contemporary social anthropologists have followed Lévi-Strauss's lead, but Chris Knight has tried to show in his book *Blood Relations* that the symbolism used by traditional peoples is often designed to make sense of some particularly difficult or important natural phenomenon.

Robin Horton has similarly defended the view that one important function of religion for traditional peoples is that it acts as a form of (scientific) explanation for the phenomena they experience (but can't explain in simple physical terms). Religion has the important function of giving coherence to the world and making sense of it. The important point here is, surely, that any kind of theory that makes sense of the correlations you observe and gives them coherence is quite good enough for the purposes of everyday life. If I happen to believe that thunder is the god Zeus stomping off to the celestial bathroom to relieve himself, then not only is that in principle a perfectly reasonable scientific hypothesis but it is also one that functions well at the cognitive level: it makes it easy for me to understand the correlation between the thunder and the rain that follows, and so ensures that when I hear thunder I head for cover with alacrity. For all practical purposes, the technological effort required to test that hypothesis isn't worth it.

In fact Horton argues that we can make even stronger claims for the scientific behaviour of traditional peoples. The fact that African witch-doctors search for reasons why their incantations have not worked is, he suggests, evidence that they really are trying to control or influence events in the natural world, the claims of his Durkheim-influenced cultural relativist colleagues notwithstanding. He quotes examples noted in passing by other anthropologists where witch-doctors have changed their incantations or preferred gods after a string of failures, and asks what is this if it isn't a form of hypothesis-testing?

The extent to which different cultures do in fact converge on the same answer when they seriously set about trying to solve the same problem is striking. In his book *The Crest of the Peacock*, George Joseph demonstrates that the Babylonians, the Greeks and the Chinese all produced workable solutions to the problem we know today as Pythagoras's Theorem (the rule for determining the length of the hypotenuse on a right-angled triangle). They used different methods to tackle the problem, but achieved solutions of similar accuracy. Similarly they and the ancient Egyptians all produced estimates of the irrational number π (without which the geometry of circles is impossible). In all these cases the discovery of the solution was dictated by the pressing need to deal with everyday problems of architectural design, either for the construction of religious buildings or the partitioning of fields among inheritors. The fact that the solutions had to be seen to be workable meant that mathematicians' ideas were under an intense form of (non-genetic!) Darwinian selection.

I shall return to this issue in Chapter 5 because it raises the question of why we seem to think naturally in this way. First, however, I want to take the argument one step further by asking whether science, in the sense implied by the examples discussed in this chapter, is something peculiar to human beings. What evidence is there to suggest that low-level empirical science is practised by non-human animals?

4
The Roots of Science

The world is independent of my will.
Ludwig Wittgenstein: *Tractatus Logico-Philosophicus* (1922)

I have argued that empirical science is a genuine human universal. In this chapter, I shall broaden this claim even further by arguing that the scientific method is not merely typical of all humans, but is also a key feature in the lives of most birds and mammals. Science as we know it in the Western world is the product of a highly formalized version of something very basic to life, namely the business of learning about regularities in the world. Being able to predict what is going to happen in order to be able to act in an appropriate way at the right moment is fundamental to survival. Organisms that wait until something has happened before responding to it tend not to survive for very long. Sooner or later they will be caught with their ecological pants down.

It seems to me that two processes underpin any organism's ability to learn about the world: these are classification and causal inference. The first provides the generalizations that allow us to note regularities in the world, while the second allows us to make sense of these regularities by organizing them into coherent logical sequences. This second process has an additional and quite important function: it allows us to store information in an extremely efficient way. Forming hypotheses about the causal relationships between the phenomena we observe allows us to reconstruct past events whose details we have forgotten. It also allows us to reconstruct the likely course of events in the future in ways that are difficult to do if all we have are generalizations. Only if we can infer causal hypotheses can we build up logical chains of cause–effect sequences that allow us to make leaps of inference from one state of affairs to another. So the question that confronts us now is: to what extent do other species exhibit these fundamental scientific abilities?

Nature's Own Scientists

From about the 1920s until a decade or so ago, scientists were reluctant to believe that animals have mental states, that they 'think' about what they are doing in the way that humans are supposed to do. Rather, it was argued that they learn simple behavioural rules (such as 'This situation is associated with a painful/pleasurable stimulus, therefore avoid/approach it'). The animal probably does not understand *why* a particular set of events should occur together, merely that a given situation is associated in some obscure way with pain or pleasure. It has simply learned a behavioural rule in response to an environmental cue. However, evidence obtained during the last decade or so suggests that animals do possess the ability to form generalizations and simple causal hypotheses.

Richard Herrnstein carried out an extensive series of studies on pigeons' abilities to form abstract concepts. The experimental design is very simple. A pigeon is presented with two tiny screens on to which are projected transparencies. Pecking at one of the screens produces a food reward, while pecking at the other produces nothing. In each case, a slide showing an example of a particular concept is projected on to the rewarded screen, and a slide of something entirely different is projected on to the other one. The rewarded and unrewarded screens are, of course, alternated at random to prevent the pigeon inadvertently learning the wrong response rule (such as 'Always peck the right-hand screen, irrespective of what's on it'). With only two screens to choose from, we would expect a pigeon that chose screens at random to hit the correct one on about half the occasions. To do substantially better than this, the pigeon must learn the correct generalization (i.e. a concept).

One example of the type of concept being taught is the concept *tree* (or *tree-ness*). During the training sessions, the pigeon would be shown hundreds of different examples of the concept: these might include particular trees of different species, aerial views of forests, forests in close-up, branches of trees, woods in winter, woods in summer and so on. These would be paired with things that are not examples of trees, such as wooden fences, wooden houses, electricity pylons and flag poles. Once the pigeon had learned to respond reliably by always pecking at the *tree* screen, it would then be tested with an entirely new set of slides it had never seen before that showed different examples of the same concept. The accuracy with which pigeons perform on these tests is quite remarkable. Although there is some debate as to exactly how pigeons manage to do this, it is clear that they are able to make generalizations that will function quite adequately as concepts in an everyday sense. They might not count as abstract concepts in the sense that you and I could sit down and decide on a formal dictionary definition of the concept *tree* that would allow us to identify new trees by ticking off a list of key characteristics. But they probably *are* doing much the same as you and I do when, in the less formal world of everyday life, we instantly know what someone else means when they assert that 'An oak is a kind of tree.'

What is significant about these experiments is the fact that they have been done on pigeons. Pigeons are not generally considered to be among the most formidable intellectual giants in the bird world. So if these animals can perform to such high standards, we should not be surprised to find that species as advanced as primates can do even better. And, as we might expect, primates generally do perform better on cognitively more complex tasks. One of the standard tests used by psychologists to assess an animal's ability to learn an abstract rule is whether it can generalize its learned response rule ('Press the lever below pictures of type A, but not those below pictures of type B, in order to get a food reward') to other examples of the same general kind. The Japanese psychologist Tetsuro Matsuzawa and his colleagues have recently shown that pigeons can learn a concept such as 'human being' (in contrast to 'non-human') quite readily and will subsequently generalize their response to some simple transformations of the concept (such as distant human faces in contrast to non-human faces); however, they failed both to transfer the learned response to other kinds of transformation (such as distant human faces as opposed to ones seen close-up) or to other types of comparison (e.g. monkey versus non-monkey or pigeon versus non-pigeon). Monkeys, while poorer at transferring to the first type of transformation (possibly due to their less acute vision), readily transferred their learned responses to transformations of the second kind. This may be because pigeons learn to discriminate stimuli on the basis of more specific features, whereas monkeys learn to recognize more general features. Why this should be so is at present unclear, though it is likely to be related to the fact that these two species have very different lifestyles which impose different kinds of demands on their information-gathering skills.

My second example also concerns a primate. During the past couple of decades, considerable effort has been invested into teaching artificial languages to chimpanzees (in particular). In more recent years the main burden of this research has focused on the use of language as a tool for exploring the animals' mental world. Work on colour-naming by Matsuzawa, for example, suggests that the way in which chimpanzees see the world is indeed very similar to the way that humans do, at least in the sense that they can be taught to use the same colour words that we do. He first trained his chimps to identify a particular symbol (or 'word') with a colour card of a specific wavelength in the colour spectrum. Once the chimps had learned this task, he then tested them using cards with the full range of hues. When Matsuzawa mapped the colour words used by his language-trained chimps on to the colour spectrum, the resulting distribution of colour terms matched exactly that obtained for humans by the anthropologists Berlin and Kaye some years before.

Just as with humans, chimps were likely to make arbitrary choices with hues that were intermediate between the reference colours, but they never used an entirely inappropriate colour word. This is a very conspicuous feature of colour-naming in all human cultures. In their original study, Berlin and Kaye showed that the words different cultures used to describe particular hues tended to congregate around the same point. Terms for blue-like colours tended to cluster around the same point of the colour spectrum. Some languages lack certain colour words altogether, however. Thus, some South Indian languages do not distinguish between purple and green; similarly, many West African languages do not distinguish between blue and green, while the same word is used for red and brown in the East African language Kiswahili. Anthropologists have often claimed that these findings prove that different cultures view the world in different ways. Unfortunately, this inference is a consequence of not knowing anything about the physics of colour. The important point about all these cases is that the colours concerned are always adjacent to each other on the colour hue chart: red is immediately next to brown, purple to green and green to blue. No one ever confuses black with green, or red with purple in this way.

The reason why some cultures lack certain colour words is simply a consequence of the fact that language forces us to categorize into discrete chunks what is, in reality, a continuum. Despite the eulogies on its behalf, language is in fact a surprisingly poor means of communication about the natural world. Because we find it impossibly difficult to discuss the natural phenomena of the world as continua, we break them up into discrete categories. When discrete categories are imposed on a continuum, the boundaries will inevitably be somewhat arbitrary. This means that some languages will call blue what we might call green when the particular colour (say, aquamarine) is on the borderland between the two 'pure' colours, even though we both agree that green is green and blue is blue when it comes to hues in the middle of each colour's range. This is simply a consequence of the way our colour detection machinery is designed.

A more significant example of monkeys' abilities to categorize the world is provided by a series of experiments carried out by Verena Dasser. She presented macaque monkeys with a discrimination task in which they had to press the lever corresponding to a photograph of a close relative in preference to a photograph of an unrelated animal from the same group. Having learned this task, the monkeys were then presented with a more complex version in which they were presented with a photograph of another group member and then asked whether they could identify *that* individual's relative. The fact that the monkeys could solve this discrimination task successfully means that they can make the judgement 'Jim is to Florence as my brother Peter is to me.' Quite obviously, the monkeys did not think in terms of genetic relatedness in these cases: lacking a theory of genetics, they could not possibly distinguish between cousins and nephews, for example. Instead, they were probably using a criterion based on the frequency of association or grooming ('Jim grooms a lot with Flo' or 'Jim often acts as Flo's ally in fights'). It seems that monkeys and apes classify the individual members of their social groups into classes or types, with these types sometimes being based on more abstract concepts. Dasser's colleague Hans Sigg, for example, was able to show in another series of experiments that the monkeys understood the relationship 'subordinate–dominant'.

David Premack has reinforced these findings in an extensive series of experiments on language-trained chimps. The chimps had been trained to use a symbolic language in which coloured plastic shapes stood for particular words or concepts which could then be used to construct simple sentences. Once the chimps were fluent in this language, they were asked to make 'same/different' judgements by identifying objects (or pictures of objects) that bore the same relationship to each other. Thus, when shown the words for a pair of scissors and a piece of paper as the target items (two items linked by a functional relationship), they were then shown, say, the word for a can-opener and offered the words for a series of possible objects that might bear an analogous functional relationship to it (e.g. a tin can, a piece of string or a paper-weight). This is just the same kind of reasoning task that is often set in intelligence tests: 'Scissors are to paper as tin-opener is to . . .?' Premack's chimpanzees were able to make the correct response, indicating that they could work out the abstract relationship '*is-related-to*'.

Learning to classify the world into categories may seem like a trivial task, but it is fundamentally important both in everyday life and in science: it is what allows us to make generalizations. However, science, as we have seen, does not consist simply of inductive generalizations in the way many eighteenth- and nineteenth-century philosophers supposed. It also consists of explanations for those empirical generalizations. Fundamental to this process is the attribution of causal relationships of the form 'Y always follows X *because* . . . (*of some cause–effect relationship*).' The question we really need to ask at this point is this: Is there any evidence to suggest that animals also use a cause–effect approach as the basis for describing the relationship between two events in the world as they experience it?

In his book *Contemporary Animal Learning Theory*, psychologist Tony Dickinson points out that there are two ways animals might store information about their experiences. According to the traditional behaviourist view, they can only do so in the form of behavioural instructions: 'If X happens, do Y.' More recently, however, some psychologists have begun to adopt a more cognitive (or mentalistic) view which assumes that animals can reason about events in the world. In this case, they must be able to store information about the world in the form of causal hypotheses: 'If X happens, then Z will follow, so do Y.' The difference is crucial, because only in the second case can the animal use the rules of logic to make a 'leap of inference' by linking two sets of correlations to infer a third relationship. Let me explain how it works.

Suppose an animal is trained on two separate association problems, so that it learns, first, that 'whenever A occurs B will follow' (making X an appropriate behavioural response), and then second, that 'whenever B occurs C will follow' (making Y an appropriate response). By storing information as behavioural instructions, the animal would have learned only two unrelated behavioural rules ('If A, do X; if C, do Y') that have no common term that would allow it to make an inference from event A to action Y. However, if it stores information as hypotheses, it would learn the simple causal statements 'If A, then B; therefore do X' and 'If B, then C; therefore do Y', from which it could easily infer 'If A, then B, and hence C; therefore do Y.'

What we have here are two hypotheses about how animals 'think' about the world that yield mutually exclusive predictions about their abilities to solve a particular kind of problem. If knowledge is stored as hypotheses, animals will be able to solve a problem correctly using inferential reasoning; but if knowledge is stored as behavioural instructions, they will not be able to do so.

As it happens, rats have been tested on precisely this problem. Holland and Straub trained rats to associate the occurrence of a flashing light with the subsequent arrival of food down a shute. Later, they placed some of the same food in the cage and allowed the rats to associate the simple act of eating with mild nausea (induced by an injection of otherwise harmless lithium chloride) in the absence of any warning lights. If rats store knowledge as behavioural sequences, their knowledge will consist of two sets of instructions ('When the light comes on, approach the food hopper' and 'If you see this food, don't eat it'), but the form in which the instructions are coded means that there is no common term that can be used to make an inference about the relationship between approaching the hopper and feeling nauseous. This is because the only possible middle term ('When you approach the hopper, food will appear') is not a behavioural instruction that the rat can learn. Hence if the behaviourists are right, the rats should continue to approach the hopper when they are retested with the original flashing light because they will fail to connect the light with the feelings of nausea that follow from eating the food. In contrast, if the cognitivists are right about how the animals store knowledge, then the rats should refuse to approach the food hopper because they *can* make the inferential leap from the two causal hypotheses 'the light causes food (to appear)' and 'eating this food causes nausea' to draw the conclusion 'if the light comes on . . . you will feel nauseous, so stay away from the hopper.'

The results obtained by Holland and Straub were quite unequivocal. The rats significantly reduced the frequency with which they approached the food hopper when they were presented only with the flashing light in the subsequent retest trials. In contrast, a control group who were trained on the first stage (light plus food) but not the second stage continued to approach the hopper at the same rate as they did before. It seems that rats, at least, store knowledge about the world in the form of simple causal hypotheses. Doubtless these are not as sophisticated as the causal hypotheses used by humans, and we might not be able to translate them directly into English. But they clearly use some form of causal reasoning that can be interpreted in the same all-purpose 'meta-language' (such as symbolic logic) as we could use to interpret the equivalent English sentences.

It is important to notice in this context that the causal hypotheses that the rats appear to be using need only take the form of simple correlations: 'Y follows X.' They need not have any *explanation* (or theory) in mind as to *why* X is always followed by Y. In this respect, the rats conform to the classic definition of causation given by David Hume, the eighteenth-century Scots philosopher. Hume had asked what we mean when we say 'X causes Y' and suggested that all we can in fact be saying is that 'X is associated very closely in time and space with Y.' He argued very cogently for the view that *all* causal statements (even in science) are of just this kind.

Now it is possible to interpret Hume's argument here in two different ways. One is to say that statements about causes are simply statements about the correlation between two particular phenomena: the billiard ball always moves when the cue is moved towards the ball (but I have absolutely no idea why it does so). The other is to say that the ball moves in the way it does *because* energy is imparted to it when the cue strikes it (in a way that does not happen if the cue actually fails to connect with the ball – *even though* cue and ball are closely associated in time and space). In the first case, we simply recognize a correlation; in the second, we offer an explanation about why the correlation takes the precise form it does. This is, of course, no more and no less than the distinction between cookbook science and explanatory science that I made in Chapter 2.

At this point, we can say nothing about how the rat construes the nature of the causal relationship between the onset of the light and the subsequent feeling of nausea. To count as cookbook science, however, it is enough that the rat recognizes that a temporal correlation exists between the two events. Of course, it is perfectly *possible* that the rat has some kind of explanation for the correlation. After all, it is difficult to place any other interpretation on the animal's experience of the effect of its own actions on objects. Whilst it may not understand why thunder clouds always presage rain, it surely must understand that, if it touches a ball with its paw, the ball will roll away, and hence that the ball moves *because* its paw pushed it. If it cannot learn the significance of this, I cannot understand how the rat can possibly survive in an environment that is as full of obstacles as the real world is.

At the present time we have no way of finding out how much an animal can infer about causal relationships. However, to count as a genuine case of explanatory science, any hypothetical explanation that is compatible with the spatio-temporal correlation between two events would be perfectly acceptable. The explanation it produces does not have to be identical (or even similar) to Newtonian physics. It would still count as explanatory science if the rat believed that the billiard ball was in actual fact a rat-like animal that lived in a green baize world and always ran away when tormented by wicked stick animals. We obviously would not want to claim that the rat's understanding of the world was particularly sophisticated (or indeed, even correct), but that is neither here nor there. It is our privilege to have access to kinds of information that are denied the rat, and so to build more complex (realistic?) theories. We do need to beware of confusing the ability to think causally with the production of particular types of causal hypotheses, otherwise we will once again be forced to assert that Newton wasn't a scientist because his theory of gravity turned out to be wrong in certain key respects.

It is possible, however, that animals like rats do learn 'natural' causal relationships – that is to say, they come to essentially the same conclusions about certain kinds of causal relationships as we do. Some evidence suggesting that this might be so comes from a series of experiments carried out by psychologists Michael Domjam and Nancy Wilson. These experiments suggest that rats genuinely do have simple expectations (i.e. theories) about how the world works, and that these expectations are based on their experience of the world. The experiments involved pairing either nausea or a mild electric shock (of the kind one might get from the electric fences that farmers use to prevent sheep and cows straying from a field) with either the delivery of saccharin liquid or the sound of a buzzer in all possible combinations. The rats found it significantly easier to associate nausea with a drink of saccharin water and the shocks with the buzzer than they did an association between nausea and the buzzer or the electric shock with the drink. Rats, like humans, have extensive experience of the fact that food and drink sometimes make you ill, and laboratory rats, at least, probably have as much experience of the fact that buzzers are commonly associated with mild electric shocks since this is such a common laboratory procedure. It seems that the rats found combinations that were outside their normal everyday experience much harder to learn. It is as though the causal linkages are more difficult to recognize in these cases. Like Sherlock Holmes, they look for plausible causal relationships first and only resort to the implausible ones once they have worked their way through the plausible ones and found them wanting. What this implies is that, in effect, rats have a theory of biology.

Once again we need to take care not to be confused by the various layers of explanation involved here. This does not mean to say that rats have the *same* theory of biology as we do; merely that they have theories of the same *general* kind. Just as we would not expect two different cultures to produce exactly the same theory to explain why the sun rises every day (even though they both agree that the sun does rise every day), so there is no reason to expect different species to do so. The theories that two different species or cultures arrive at to explain a particular phenomenon will depend partly on the kinds of experience each has had of the phenomenon, partly on the importance that the phenomenon has for them (and hence on the effort they are prepared to put into finding out about it) and partly on their cognitive skills (irrespective of whether these are genetically determined species differences or learned cultural differences). They will also depend on the individuals' respective technological skills: the kinds of explanations you can aspire to are inevitably limited by the kinds of technology you have for exploring the natural world.

Children as Natural Scientists

We can come slightly closer to home by asking about children's abilities in this respect. Young children are in no sense trained scientists, and the behaviour of pre-school children in particular can tell us a great deal about the natural abilities of humans in the period before adults have had time to instil any culture-specific patterns of behaviour.

It is, of course, quite obvious that children learn. Recognizing this, we devote considerable amounts of time and money to providing them with the right environment in which to learn. This has given rise to the belief that children are like sponges that soak up culturally determined knowledge. Children's abilities to absorb knowledge and beliefs are, of course, legendary, but they can easily distract us from children's equally remarkable abilities to work things out for themselves.

In recent years psychologists have been particularly interested in how children learn about each other's minds – how they make the transition from the inanimate to the animate world. Once children realize that some of the objects in their perceived world are sentient beings that are capable of moving objects around, they enter an entirely different and very much more sophisticated phase of existence. From this point on they are aware that other individuals can be cajoled, manipulated or even deceived into behaving in ways that are to the child's advantage, to give the child things it does not have or cannot reach for itself.

Paul Harris has argued that although children develop this ability quite naturally during the first few years of life, it does not develop as an inevitable unfolding of the genetic programme with which every individual is endowed. Rather, children arrive at it by a gradual process of understanding based on experience and deliberate attempts to simulate or model the world in order to control it better.

A major component of this process involves the development of pretence, especially in the form of fictional play. Harris argues that children begin life with a number of default settings that automatically guide its perceptions of the world. Pretence, and then the ability to attribute mental states to others, develops as a series of stages as the child learns more about its environment. It does not need to be told about these; rather, by comparing its own mental experiences with its perceptions of other individuals' behaviour, it gradually builds up a series of models of increasing complexity.

This series of transitions is exemplified by the child's attitudes towards dolls. Initially these are treated as passive recipients of the child's attentions; later, the child begins to treat the doll as being an active agent (who can, for example, pour tea at the dolls' tea party), then as having sensations and perceptions and finally as being capable of holding false beliefs about the world. Young children will squeal with delight when Punch threatens to throw the coffin over the edge of the theatre because they know that Punch believes Judy is still in the box whereas *they* know she escaped from it while Punch's back was turned (i.e. Punch has a false belief about something). Only much later do they come to realize that dolls do not behave like people, and they then begin to make a distinction between the cognitive attributes of dolls and people. In effect, Harris argues, children are running a series of simulations about the nature of the world. At each stage they hold certain beliefs about what the things they see are capable of doing, and they modify these as their knowledge of these objects grows. In other words, they learn by experience, but they use a set of in-built preconceptions to guide their investigations.

Harris specifically draws attention to the parallel between children and scientists in this respect. He focuses in particular on the way children use analogies based on their growing understanding of their own minds as a basis for making predictions about other individuals. Like the scientist's initial hypotheses about a new phenomenon, the child's first analogies are drawn from its own internal world. As it tests the predictions of these hypotheses against the actual behaviour of the objects to which they refer, it gradually refines its hypotheses in the light of the evidence.

This process seems much in evidence when children learn languages. Although psychologists disagree as to whether children initially learn meanings for words that are narrower or broader than those used by adults, they nonetheless agree that children converge on the correct meaning (i.e. the one broadly accepted by the adults in their linguistic community) by a process of hypothesis-testing. In effect the child initially infers that a word has a certain meaning from the context in which it first hears it being used. Subsequently it refines that definition (i.e. hypothesis) as it hears the same word being used in other novel contexts (so allowing it to eliminate irrelevant cues) or when its own inappropriate use of the word is corrected by an adult.

Hypothesis-testing is also very much in evidence in the way children learn word endings (for example, the correct form of the past tenses of English verbs). In an exhaustive review of this aspect of language-acquisition, psychologist Eve Clark has argued that even though children may initially use examples learned by rote as models, they soon begin to learn general rules about how endings are normally constructed (e.g. 'Add *-ed* to the end of the verb'). Rules of this kind are, to all intents and purposes, hypotheses about how the world works.

Notice here that the rule that the psychologist attributes to the child ('Add *-ed* to the end of the verb') is not what the child itself actually has in mind, any more than the rat says to itself 'If the buzzer sounds, that means food will be delivered down the shute.' In both cases the rules are probably held in some kind of iconic or other non-verbal form. Nonetheless we can represent that form quite adequately in words: it is simply a form of translation from one kind of language to another.

Other aspects of the way children learn languages are reminiscent of some of the rat experiments I described earlier. There is evidence to suggest, for example, that children learn words more easily for categories of objects that they are familiar with (such as toys) than they do words for categories that do not occur as natural objects (for example, unicorns). This might reflect either the fact that children naturally learn to categorize the objects they encounter into types based on similarity, or that there are innate mechanisms that naturally 'scaffold' children's acquisition of knowledge about the world by predisposing them to pay attention to certain features of the world. Although this issue is far from resolved, children do seem to attend naturally to certain kinds of cues long before they can possibly have been influenced by language.

The notion of agency appears to be an integral part of the way children see the world. They naturally place the subject before the verb when they first start to use languages (at least in those languages like English where subjects stand alone rather than being specified only as word endings). Robertson and Suci have shown that children as young as eighteen months old consistently attend to the agent when shown a film clip of an agent–action–recipient sequence. Much of this seems to be related to the child's acquisition of a 'theory of mind' (the ability to infer another individual's state of mind). Children as young as eight months of age, for example, stare longer at a picture when the agent in the action is a person than when it is an inanimate object.

Direct evidence that very young children interpret events in the world in terms of causal relationships has been obtained from a series of experiments by Alan Leslie. Using a format in which young infants are shown short filmed sequences, he found that children as young as four-and-a-half months are more sensitive to film clips showing 'natural' causal relationships between objects (e.g. a green brick moving immediately after being struck by a red brick) than to 'unnatural' relationships (e.g. the green brick moving after the red brick came to a halt 6 cm short of it, or a half-second after being struck by the red brick). Moreover, by six months of age infants showed more recovery of interest (measured in terms of the length of attention) to a reversal of the sequence after habituating to the natural sequence than to the unnatural one. Since the spatio-temporal relationship between the bricks is the same in the 'normal' and 'reversed' sequences, this suggests that infants perceive something other than just spatial continuity between the two stimuli.

Leslie argues that these results undermine Hume's assertion that our perceptions of causality are based on generalizations from the repeated experience of the co-occurrence of two objects or events. Instead, it seems to offer some support for the claim that causal relationships are perceived directly. In other words, when we see conjunctions of this kind we naturally tend to interpret them in terms of causation. Leslie argues that very young children are especially sensitive to these kinds of spatio-temporal conjunction in their first year of life, and that this sensitivity provides the basis from which an understanding of causality later develops. Recent experimental studies have shown that by the age of three years children are already making use of quite sophisticated notions of causality in their attempts to understand simple mechanical interactions.

It is, of course, always possible to argue that even children as young as four months of age have had plenty of time in which to learn about causal relationships, perhaps through interactions with their carers. There is little to be gained by getting embroiled in a fruitless nature-versus-nurture argument on the origins of a causal sense. Whether it is in some sense 'innate' or whether it is learned (or more plausibly, perhaps, a combination of the two) is immaterial to the main point that very young children already have a genuine understanding of causality long before they learn to speak. In other words, children begin to interpret events in the world in causal terms long before they can possibly have been influenced by the cultural preconceptions of the society into which they happen to have been born.

Leslie has taken this argument even further by suggesting that this causal sense is a crucial component in the child's development: it provides a kind of scaffolding that automatically gives the world sufficient coherence for the child to begin to make sense of its real complexities. He argues that it is this notion of natural causality that, within two to three years, allows the child to develop the ability to engage in pretend play that involves objects that do not actually exist.

Learning Rules of Thumb

Biologists have learned to make a clear distinction between what scientists are able to infer about the factors guiding an animal's behaviour and the rules that the animal itself uses. Because we (as scientists) can stand outside of the animal's own world, we can appreciate the long-term functional consequences of what it does and so determine how evolutionary forces mould the animal's behaviour. The animal itself cannot do this, for it works at the much more proximate level of motivations. It uses simple rules based on a combination of past experience and motivation: it feels hungry, so it goes to where it knows it can find food. Biologists refer to these as 'rules of thumb'.

During the process of learning from its experiences, an animal develops simple rules to guide its behaviour that are based on generalizations. We, of course, often do the same: when we learn not to put our hands into the fire because it hurts, we learn a simple (if painful) rule and use it to guide our future behaviour.

Let me return to the honeyguide and the Boran pastoralist whom we met earlier. In that context, I used this case as an example of how a traditional tribesman, acting as a natural scientist, can learn from experience how to exploit the behaviour of the natural world in which he lives. Let me now reverse the problem and ask what the bird can tell us about animals' abilities to act as scientists.

Although it is possible that honeyguides learned their guiding behaviour from humans (records of this behaviour being exploited by humans exist from as long ago as the seventeenth century), a more likely explanation is that the bird has generalized this behaviour to humans from other species. This is suggested by the fact that the birds guide ratels (otherwise known as the 'honey badger') to bees' nests throughout this species' range in Africa. The bird uses exactly the same behaviour to attract and guide the ratel as it does with humans. Raiding bees' nests to eat the honeycomb and grubs is one of the ratel's favourite activities. Its thick skin and dense fur make it almost impervious to bee stings, these probably being adaptations that evolved to allow the ratel to break open bees' nests. Although the ratel probably does not deliberately leave honeycomb for the honeyguide in the way the humans do, it is a rather messy eater and invariably leaves enough to make guiding worth the bird's while.

What is interesting about the honeyguide's behaviour is the way it fine-tunes its actions to the responses of the individual it is trying to guide to a nest. Isack and Reyer found that the honeyguide monitors the human's behaviour very carefully. By alternately flying off towards the nest and then waiting for the following human to catch up (or returning to collect him if he does not), the bird guides the man progressively towards the nest. Once near the nest, the bird changes its behaviour: it gives a different call (the 'indicator' call) and is less responsive to the shouts or noises made by the following human. As soon as the human has approached, it will fly only a short distance further on before stopping again, finally flying in circles around the nest once it reaches it.

Even more interesting, perhaps, is a claim made by the Boran honey-gatherers that the birds will attempt to deceive them by deliberately underestimating the distance to the bees' nest when it is more than about 2 km (approximately the limit to which Boran are prepared to go in search of a nest because of the length of time for which their herds will be left unguarded). Isack and Reyer were unable to put this claim to the test; however, given that they had substantiated all the other claims made by the Boran, they felt that this one could well be true too.

One obvious alternative explanation for this last claim is, of course, that the birds are just not very good at estimating longer distances. However, if this were the case, we would have to explain why the birds are so good at estimating distances only up to those to which the Boran are prepared to follow them, yet so poor at longer distances. This can hardly be a coincidence. If, on the other hand, the Boran are right in their claim that the birds deliberately underestimate long distances, then it suggests that the birds are aware that humans are unlikely to follow them to nests that are too far away.

This implies extraordinary sensitivity on the bird's part to the behaviour of another species. But it is not a feat beyond the powers of many of the advanced animals. As Dorothy Cheney and Robert Seyfarth have demonstrated so elegantly in a long series of field experiments, vervet monkeys (at least) are very good ethologists in the sense that they can read other individuals' intentions from their behaviour extremely well. Indeed, their work in itself counts as good experimental evidence that animals are able to make simple 'If . . ., then . . .' inferences, for they have shown that these monkeys can use information about the past behaviour of animals both to choose the best coalition partners and to exploit and deceive other members of their own species.

Another example of natural science concerns the use of medicinal plants by chimpanzees. There are now several well-documented cases of chimpanzees using natural plant products that are known to have medicinal properties. Mike Huffman and Mohammedi Seifu noticed that the chimpanzees of Tanzania's Mahale Mountains National Park sometimes chew the stems of *Vernonia*, a common East African bush, but spit the chewed pith out. The juice from the stems of this plant is very bitter and rather unpalatable, but it is used by the local people of this region to treat intestinal disorders (especially parasites). Huffman and his colleagues have since been able to show that chimpanzees are more likely to eat *Vernonia* if they have intestinal parasites, and that after consuming the plant their parasite load is significantly reduced. It is difficult to see why chimpanzees should go to the length of chewing (but not swallowing) stems of this bush had they not discovered that the juices have some benefit.

Similarly Toshisada Nishida and Paul Newton noticed that the chimpanzees sometimes roll the chewed leaves of the herb *Aspilia* around their lower lips for some time before spitting out the residue. *Aspilia* is also used as a herbal medicine by the local people in the treatment of intestinal disorders. Nishida and Newton point out (and Paul Newton is both a medical doctor and a professional primatologist) that the gum is a particularly absorbant surface. Medicines can be absorbed through the gum directly into the bloodstream under circumstances where the compounds concerned would be destroyed in the stomach by the digestive juices.

Chimpanzees are not the only species to have noted the correlation between health and particular foods. Gelada baboons living in the Simen Mountains in northern Ethiopia face problems in the seasonal availability of their preferred foods (grass leaves). During the dry season, the grasses dessicate badly in the sun, and become very difficult to digest. During my own field work on this species, I noticed that the animals became seriously constipated at this time of year. Towards the end of the dry season, however, the wild roses come into fruit. The animals compete intensely for access to rosehips and consume them in large quantities. (Indeed, they even compete with the local people for these fruits.) From the moment they start to eat rosehips, the animals cease to suffer from constipation, despite the fact that the grasses on which they are feeding won't become lush and soft until the rains break a month or so later. Rosehips have a high vitamin C content, and vitamin C is an effective cure for constipation.

My last example, once again, concerns rats. Animals' abilities to make inferences about the world in which they live depends, in the final analysis, on their ability to acquire information about the regularities in the world's behaviour and, from these, to infer general rules of thumb that allow them to predict with tolerable accuracy what is likely to happen in the future. A particularly enlightening set of experiments on just this point has been carried out by Tony Dickinson and his colleagues. In these experiments, rats were trained to learn a simple rule of thumb for solving a maze problem in which the animals had to turn either right or left down a simple T-shaped maze to find a food reward. The experimenters were interested in the effects of practice on the animals' abilities to solve problems. Two groups of rats were allowed different amounts of practice on this problem, and then tested on the reversed rule. In other words, where they previously had to turn right to find the food, they now had to turn left. Rats that received *less* practice on the problem learned the new response rule quicker than those that had been allowed more practice.

Dickinson argues that the only plausible interpretation of these results is that when the rats encounter a new problem, they have to think about it; but once they have learned what the correct response is, the rule is transferred into a habit that they don't have to think about. This allows the animal to call up the appropriate rule automatically whenever it finds itself in the same situation, thereby avoiding the need to waste too much time thinking about what it should do. But once a rule has been learned, increased practice turns it into a habit, and it then becomes more difficult to unlearn.

It is clear that this argument bears many parallels to what happens in humans under comparable circumstances. When we first learn to drive a car, we have to concentrate very hard on what to do. However, once the behaviour sequences involved become automatic, we no longer have to think about them consciously. Instead, we can drive while holding a conversation, listening to the radio or admiring the scenery. But the moment there is a major change in the routines (as happens, for example, when we cross over to the European continent after driving in England, and now have to drive on the other side of the road), we are forced to think very carefully about each and every move. We find it harder to do the right thing at the right time and can no longer rely on automatic pilot.

In the previous chapter I argued that many of the features that characterize science are also to be found in the daily lives of other 'pre-scientific' human societies. In this chapter, I extended this claim to other species of animals. The main emphasis here, of course, has been on cookbook science. My concern has been to show that the basic processes that underlie science are neither something especially unusual nor peculiar to one particular culture. Rather, science is a genuine universal, characteristic of all advanced life-forms.

The essence of my argument has been that empirical science is something intrinsic to life itself for these organisms. The ability to engage in empirical science makes it possible for these species of animals to operate much more effectively in the world: it allows them to respond flexibly to situations as they occur in conditions where the future is generally unpredictable. In short, science really is just plain simple learning of the kind with which we are all familiar. Claims that other species or cultures do not engage in science are largely a consequence of confusing the nature of science as a process with the particular theories that certain individuals have arrived at as a result of applying this process.

In the next chapter I ask why this method of finding out about the world should be so successful.

5
Why is Science so Successful?

If at any future time [the facts] are ascertained, then credence should be given to the direct evidence of the senses rather than the theories.

Aristotle: *De Generatione Animalium* (c.330 BC)

I have argued thus far that science (that is to say, the scientific method) is not only universal among humans but that it is a 'natural' approach to the physical world in which we live in the sense that it is characteristic of all higher organisms. This is not to say that *everything* a human being does is empirical or that it is necessarily done with the analytical rigour that professional scientists have come to expect. Rather, we resort to simple rules of thumb based on generalizations that seem to work well enough to get us by. Only when our rules of thumb no longer work do we engage in the often difficult business of empirical science.

All this begs the question as to why the scientific method should be so widely adopted. What is it about the empirical approach that makes it the preferred solution?

Pragmatic Realism

As we noted in Chapter 2, one of the central philosophical problems for the empirical sciences has always been the problem of induction, or Hume's problem. How do we *know* that our theories about the world are true? Defenders of the realist view of science have argued that science's theories genuinely do interpret a reality that is 'out there' and we know this must be so because the theories work in practice. They claim that this very fact is by far the best justification we have for the methods of empirical science. So far from adopting theories more or less at random according to some arbitrary metaphysical belief, we in fact put them to the test and accept as true (at least for the moment) whichever theory produces the best results. If conducted with sufficient care, the act of testing effectively guarantees the theory's empirical validity because only a reasonably correct theory will consistently be able to predict what actually happens.

This has to be a fairly tough test because, short of cheating, there is no way in which anyone can force the world out there to behave in accordance with his or her pet theory. You might get away with it on one occasion, but science demands that experimental results be capable of replication. I have to be able to reproduce what you claim to have found, otherwise I am under no obligation to believe you.

Part of the problem has, I think, been that philosophers have grossly overestimated the way in which science is actually done. The classic (or 'naive') Popperian view is rather austere, viewing scientists as super-robots that automatically apply rather strict criteria on whether to accept or reject hypotheses. But scientists live in the real world and they know only too well how fallible they can be when it comes to designing experiments. Popper's prescription is just too stringent for real life. If we applied it rigorously, we would soon end up with all our hypotheses falsified by data simply because most of the time our tests of hypotheses yield negative results either because we have left some crucial confounding variable out of account or because we have simply designed the experiment badly. The latter is a particularly common problem in studies of animal behaviour: an unwritten law of animal experimentation says that if there is a way of solving your carefully designed experiment other than the way you intended, an animal will find it straight away. Here, encapsulated in an apocryphal story of uncertain provenance, is an example of just the sort of thing that can happen.

After a great deal of effort, so the story goes, a famous psychologist had finally succeeded in demonstrating that rats could solve an 'odd-man-out' problem. In this particular experiment, the rat had to run down a runway past a series of small doors, on each of which was a symbol (e.g. a cross, a triangle or a circle). The doors let the rat into another chamber where it received a food reward. All the doors except one had the same symbol, and all the doors which had the same symbol were locked; only the door with the 'odd one out' could be pushed open to allow the rat to gain access to the food reward. The rat's task was to run down the runway looking at all the symbols and then make just one choice as to which door to try to push open. At the end of each trial, the symbols on the doors were changed and the rat given another go. So the rule the rat had to learn was: 'If the choice is between a set of circles and a triangle, choose the door with the triangle; if offered a set of triangles and a square on the next trial, choose the door with the square', and so on. After the first few trials, the rats behaved with uncanny accuracy, solving the problem correctly trial after trial. In effect the rats had been able to learn an abstract rule. This was a major event. Here at last was proof that animals were not just machines: they could *think*. This is the stuff of which Nobel Prizes are made! Unfortunately, at this crucial point, the scientist decided to make a film of the experiment to show during his lectures. When the film was being played back in slow motion during editing, someone noticed that what the rat was doing was running along the runway past the doors like a bat out of hell, kicking each door as it went past with its back foot; as soon as it came to a door that gave slightly when kicked, it stopped and shot through it. The symbols on the doors were completely irrelevant: the rats had found another way to solve the problem. Small wonder that rodents turn out to be the masters of the universe in *The Hitchhiker's Guide to the Galaxy*: a century of psychologists' experiments on rats amply demonstrate that rats can outwit humans any time.

Long experience with problems of this kind has made scientists less ambitious than the naive Popperian view supposes. Scientists do use a falsifiability criterion as recommended by Popper, but they use it as a criterion for identifying useful hypotheses not as a prescription for how to do science. Far more important to a scientist is another much less familiar concept of Popper's, namely the notion of a 'fair test'. Popper pointed out that when the results of an experiment turn out negative, there might be two quite different reasons why this is so. One, of course, is that the hypothesis is wrong. The other is that we have omitted some crucial confounding variable (usually because we were quite unaware of its existence) and that this confounding variable radically alters the predictions made by our hypothesis. Our hypothesis stated that the more rain you get, the better the crops will grow. But the way we designed our experiment failed to take into account the fact that planting the seeds on stony ground results in very little water being absorbed by the soil no matter how hard it rains: the result is a poor crop every year irrespective of the quantity of rain that falls. In science, the phrase 'All other things being equal . . .' is probably the single most important component of any experimental test.

The business of hypothesis-testing is neither easy nor straightforward. Nature is no respecter of persons when it comes to ill-conceived theories, and it is all too easy to fall into the GIGO ('garbage-in/garbage-out') mode of science if you don't think long and hard about exactly what you are doing and precisely how your pet theory works. But if you are honest, it is all but impossible to force nature to behave in accordance with your theory when it is not designed to do so. You might manage it once by overlooking some confounding variables, but the chances of doing so repeatedly as you generate and test further predictions derived from the original theory become smaller and smaller.

What, then, defines a good theory? In his book *The Rationality of Science*, the philosopher William Newton-Smith lists eight key features. These are: (i) observational nesting (the theory's ability to explain the successes of its predecessors); (ii) fertility (its ability to generate new ideas to guide future research); (iii) track record (its achievements in making correct predictions in the past); (iv) inter-theory support (its ability to provide additional evidence in favour of another theory); (v) smoothness (the fact that it needs few auxiliary hypotheses to explain its failures); (vi) internal consistency (that it contains few statements that lead to the acceptance of logically incompatible predictions); (vii) metaphysical compatibility (that it meshes well with our other beliefs, including our general metaphysical position); and (viii) simplicity (a version of Occam's Razor which says that, when all other things are equal, simpler theories are to be preferred, if only because they will be easier to compute). The basis of his argument is that, if we apply these criteria carefully, the growth of knowledge will proceed in a genuinely rational way.

Newton-Smith is specifically concerned to counter the anti-rationalist views advocated by Kuhn and Feyerabend, both of whom insist that shifts of theoretical paradigm are largely a matter of belief because there is no way of testing between competing theories when these cannot be directly related to each other (in technical terms, that they are 'incommensurable'). The basis of this claim is that the theoretical terms of different paradigms often mean completely different things, even though the actual term used may be the same. Thus the concept of *mass* in Newtonian physics means something rather different to what it means in Relativity physics, if only because Newton assumed that mass was constant. Similarly the term *fitness* as it is now used in evolutionary biology (where it is defined as the relative rate with which a gene is propagated) does not mean quite the same as it did when it was first used in the aftermath of the Darwinian revolution (when it meant something closer to the ability to survive in a given habitat: see Chapter 8).

Newton-Smith argues that the anti-rationalists are trying to foist much too grand a goal on to science. With the more modest aim of simply trying to explain what we see, it is much easier to counter the Kuhn–Feyerabend claim. In these terms, theories are commensurable in the sense that we can understand exactly why the old theory was wrong, even though we may not be able to translate directly from one theory into the other. In many cases, for example, the earlier theory turns out to be an approximation. Survival is a component of fitness, but it's at best only half the story; the Newtonian concept of mass works just fine so long as we remain within the scale of the solar system. We can understand why the eighteenth-century biologists made the mistake of assuming that evolution is linear and progressive, or why people assumed that the earth is flat and that the sun moves round the earth. These inferences seem 'natural' in the sense that they take the world as we experience it at face value. And for many everyday purposes, it is quite reasonable to make these assumptions: we simply cannot measure the difference between the predictions of the two theories.

On the other hand, Newton-Smith is equally critical of rationalists like Popper and Lakatos, arguing that they have also taken too strict a view of scientists at work. What scientists actually do, he insists, is something much more low-key, although it is nonetheless rational: they simply rely on the fact that their theories really do work in the sense that they predict what will happen with reasonable precision (or in cases like quantum physics with a degree of precision that can be truly mind-boggling). Newton-Smith calls this Temperate Rationalism.

This pragmatic line has, in fact, been argued by other philosophers of science. Nicholas Rescher, for example, argues that the best justification we can have for science is precisely its past success. He draws a crucial distinction between Hume's conception of the problem of induction (the fact that 'has succeeded in the past' cannot imply 'will succeed in the future' – the logical error of inferring 'all cases' from 'some cases') and what he terms the *faute de mieux* justification for continuing to rely on a successful method ('the fact that a method has succeeded in the past is the best justification for continuing to use it when we have no better alternative available').

Rescher argues that this is the only rational way to proceed if we want to live and survive in the real world. To argue otherwise is to undermine the very basis of our existence. We may not be able to offer any theoretical justification that holds a priori throughout the universe (i.e. is true by definition), but that, he argues, is a minor difficulty we can afford to live with. Our real concern lies in trying to understand the world well enough to survive in it: we have to be able to plan for tomorrow. Rescher points out that almost all the discussion over the problem of induction has focused on the *theories* that emerge from it; in contrast, he wants to draw a firm distinction between justifying particular theories and justifying the methods of science. He argues that pragmatic considerations justify the method and are the best we can ever hope for (or even need to hope for!). Once we have justified the method, that alone is sufficient to justify the theories that flow from it. And so the circularity in Hume's problem is broken.

The philosopher Nicholas Maxwell wants to go further by insisting that there are even stronger grounds for justifying modern scientific practice. He argues that a fundamental assumption underlying science is that the world is comprehensible, that it has a certain internal consistency. In other words, the phenomenal world as we experience it is not wholly chaotic and lacking in pattern: we *can* find out about it, we *can* understand how it works. Given that this is a central assumption for doing any kind of science at all, he suggests that the most rational way to find out whether or not it is true (and hence whether or not we really can do science) is to assume it to be true and see how far we get. If we find that we do obtain sensible mutually consistent theories that work, then we can be fairly sure that the assumption was indeed true and science is possible; if we fail to obtain consistent results, then we will know that the assumption is false (and hence that the whole enterprise of science is pointless). He argues that we have more chance of finding out whether or not we are right about science by proceeding in this way than by agonizing in our armchairs over logical possibilities the way so many intellectuals have tried to do.

Maxwell calls this 'aim-oriented empiricism' and contrasts it with the 'standard empiricism' of conventional philosophy of science. In standard empiricism, a theory's ability to predict the observable world is taken as the touchstone of its validity. But, as Maxwell points out, this is in fact counter-productive because no theory in science is ever 100 per cent accurate: there is always some degree of experimental or measurement error, as well as the effects of confounding variables. It is always possible to think of some completely arbitrary theory that fits the facts exactly: a simple description, for example, will always fit the facts better, though it will have told us nothing about how the world works. Instead, he argues, scientists accept a theory both because it receives some empirical support and because it provides a coherent metaphysical scheme within which to work.

The only possible recourse open to a relativist at this point is to argue that the success of a theory does not guarantee its truth, merely that it happens to predict the results we observe. But now we are in danger of collapsing into solipsism (the philosophical view that only my mind exists, and the rest of the world is simply a figment of my imagination). To this, of course, there is no answer, other than to point out that people do not behave as though this were true (there would be no reason to be generous or to help others if it were, for example).

Mental Models

We can, I think, go one step further in justifying the pragmatic arguments for the validity of science. In Chapter 4, I argued that animals use essentially the same kind of causal logic that we do. Underlying this conclusion is another important idea, namely the fact that rats appear to code their knowledge about the world in the form of hypotheses. In effect, they construct models about how they think the world actually is, and use these models to predict the future.

The importance of this observation lies in the fact that not only is this a close parallel to what scientists do, but it is also a close parallel to the way psychologists are now beginning to think that people store knowledge about their experiences. This so-called 'mental models' hypothesis owes much to the work of psychologist Philip Johnson-Laird.

What has emerged from this research is the suggestion that we store knowledge in the form of hypotheses or models rather than, as is usually supposed, in the form of pictorial representations of events as they actually happened. Think of the way in which you might recall an event to which you had been a casual witness. If you are asked to say what happened (for example, did the assailant hit the victim with a spanner or his fist), you are often unable to give the correct answer. But if you are allowed to tell the story from the beginning (which involves re-running the whole sequence, step by step), you can often give the right answer. It seems that we do it by reconstructing what actually happened from a few key events that provide anchor points. It is as though we know 'the kind of things likely to happen when . . . (X is the case)', so we simply remember the key events that will allow us to reconstruct a complex sequence with reasonable accuracy. When subsequently asked to say how the assailant hit the victim, we reconstruct what must have been the case: 'He must have hit him with a spanner because I saw blood spurting everywhere, and that simply doesn't happen if you hit someone with a fist.'

No one claims that such reconstructions are perfect: indeed, witnesses to crimes are notoriously inaccurate in their recollections (and the mental models hypothesis may offer us an explanation as to why this should be so). The point is that this way of storing knowledge is good enough to allow us to function effectively in the real world most of the time. And this seems to be the crucial point. The whole mechanism is a compromise between the need to store vast quantities of trivial detail (which would very soon overload the brain's capacity to remember information) and the need to store just enough information to allow an individual to predict what will happen next with sufficient accuracy to enable him to take appropriate action. That action might be rapid evasive action (as when you notice a No. 9 bus hurtling down the road out of control towards you) or a deft change of direction to cut off the escape of a deer. Like all biological phenomena, memory processes are compromises and hence imperfect. Indeed, it is a biological truism that evolution tends to produce solutions that are good enough to get by rather than the kinds of engineering perfection that we are used to from the physical sciences or our own technology.

Johnson-Laird's claim is that storing knowledge in the form of causal hypotheses (or models) is extremely efficient for two reasons. One is that it is doubtful whether even humans have sufficient memory capacity to learn the right response to a stimulus by induction alone. To create memory stores for every single event that might occur during a lifetime and then to remember the outcomes of every response ever made would require an enormous brain, even given the remarkable efficiency of brains compared to conventional computers. If nothing else, it would be unbelievably wasteful of memory capacity. The second point is that the phenomenal world is literally overflowing with sensory information. Most of this enormous quantity of detail is of no immediate or future relevance to anything in particular. It simply does not matter whether or not the nightingale happened to have been singing in Berkeley Square when the No. 9 bus came hurtling down the hill out of control. Either you recognize the problem instantly for what it is and dive over the wall, or you don't get a second chance to put an alternative hypothesis to the test. Nor is it especially useful under these circumstances to engage in deep philosophical debates over the validity of our knowledge of the external world and the social symbolism of large red patches spreading rapidly across the visual field. In Johnson-Laird's view, we solve the problem by learning to identify certain rather specific cues as being relevant to a particular class of problems and to ignore the others.

This is one reason, perhaps, why we seem to get locked into a particular way of viewing the world and cannot see the obvious solution for a problem until someone who can think 'laterally' points it out to us. We often then recognize the solution in a flash of realization. Science itself is often like that, as many scientists will testify. Someone publishes an important new theory and everything suddenly seems to fall into place: his/her colleagues spend the next year or so wondering why on earth they hadn't thought of it for themselves.

What we seem to do, then, is learn rules of thumb based on past experience. These rules are not necessarily perfect, but they are good enough to get us by most of the time while avoiding the immense costs (in time, if nothing else) of having to puzzle over the details of each situation. These rules are then commonly stored as habits in unconscious memory where they can be called up at a moment's notice. Only when we find that our rules of thumb are leading us astray (by making wrong predictions) do we bring the problem back into conscious memory in order to try to puzzle out a better solution. In fact we seem to do just what Dickinson's rats do. The parallel with the way scientists actually work is very striking.

These rules of thumb invariably take the form of causal hypotheses because, as the philosopher Jonathan Bennett has pointed out, thinking causally is one of the most effective ways of giving coherence to the world in which we live. This almost certainly makes it a great deal easier to make sense of the world and to predict its future state more accurately. Mental models, in other words, are naturally causal models. The anthropologist Robin Horton has made the same claim about African religious systems: they provide a model of the world that allows believers to make sense of the various things that happen in it, and so to predict the future more successfully.

Does the Scientific Programme Really Work?

Irrespective of whether we adopt a realist or an anti-realist stance, we are sooner or later confronted by the question as to whether science really has been as successful as its proponents claim. Given the kind of pragmatic position adopted by philosophers like Rescher and Newton-Smith, the relativist faces a serious challenge because his position now depends on being able to show that science is no more successful than any other belief system such as magic. In effect, this entails showing that science is successful only as often as we might expect if we selected our theories about the world at random. The difficulty here lies in trying to decide how often theories selected at random would turn out to be right by chance alone. Nonetheless it would probably be enough to show that, although we claim our current theories are more complex than those of our predecessors, our actual ability to control the world is no better.

I cannot understand how anyone can, in all honesty, claim that our ability to control events in the world has not improved over the last few centuries, let alone over the 6000 years or so since recorded history began. Such mundane things as motor cars and aeroplanes, radios and televisions, dishwashers and fridges, are everyday testimony to our ability to make the physical world work for us. An astute relativist might argue that most of these represent simple cookbook science arrived at by trial and error without the benefit of theoretical (or explanatory) science. There was a time when this was undeniably true, but in fact this is no longer the case. Modern jet engines are designed using fundamental principles of physics; the silicon chips that lie at the heart of the personal computer I am working on as I write have their circuitry etched into them using the principles of quantum physics.

To argue that these are produced by random trial-and-error learning is simply to display ignorance of how modern technology (and especially electronic technology) is designed. There surely has to be something faintly absurd about a relativist, who prepared his paper on a word-processor and travelled to an international conference on a jet aeroplane, claiming in his lecture that science is a cultural construction.

But we can give more direct proof of the validity of at least some of the theories of science. Let me offer just one example from the space exploration programme. The late 1960s found the American space programme in the middle of its widely heralded manned exploration of the moon. On 11 April 1970 the Apollo 13 mission took off from Cape Canaveral in Florida on its way to attempt yet another moon landing. However, 56 hours into the three-day flight, a fuel tank exploded, destroying much of the spacecraft's control systems as well as the rocket fuel required to bring it back to earth. The recording of the astronaut's eerily laconic message to mission control in Texas ('Hello, Houston, we have a problem here') says it all. As Houston checked out the spacecraft's systems on its giant earth-based computers, it began to dawn on the control team that the impossible had happened: a complete failure of four of the craft's main systems. Two of its three fuel cells were dead, destroyed when one of its two oxygen tanks exploded; meanwhile the contents of the other oxygen tank was bubbling away into space. Mission control was faced with a serious problem: how were they to get the astronauts back to earth? With its main rocket system all but disabled, the spacecraft had no means of turning round and blasting its way back into earth orbit. In the virtually frictionless vacuum of space, motion will continue for ever in a straight line once it has started: the three astronauts were on the verge of an unplanned perpetual mission into deepest space.

Their solution to the problem was straight out of science fiction, as well as being an extraordinary gamble on the validity of Newton's theory of gravity. Mission control decided to allow the crippled spacecraft to continue on its way to the moon; then, by a judicious firing of the rockets on the still-attached moonlander, they would use the moon's gravitational field as a slingshot to propel it back towards the earth. Although the moon's gravitational field could be used to accelerate the spacecraft out of its own influence, the timing of the rocket firings required to make this possible had to be very precise indeed if it was to break free of the moon's gravitational field at just the right point to make it back to earth. Because this would have to take place on the far side of the moon, Houston could not control the spacecraft's computers directly. Instead, the long list of instructions had to be radioed up to the astronauts to be fed into the spacecraft's computer. It took the astronauts two precious hours just to write it down, line by line, using every scrap of paper they could lay their hands on in the process. Throughout the world, people were glued to their radios and TV sets in what rapidly became the media event of the decade.

Happily, history tells us that the gamble paid off, though not without some heart-stopping moments before the astronauts splashed down in the Pacific. More importantly, the successful outcome of this story can only mean that Newtonian theories of physics work well enough at the scale of the solar system to be considered a true description of reality. After all, men's lives were being staked on what the relativists insist are quite arbitrary metaphysical beliefs. If this isn't a powerful vindication of the theories of science, then I simply do not know what is.

Modular Science

Given that science has been so successful, we have to ask why: what is it about the methods of science that has made it possible to achieve such spectacular successes? One answer, of course, is rigorous hypothesis-testing, but this is not really the answer we are looking for. After all, the burden of my argument in the preceding chapters has been that everyday life consists largely of informal hypothesis-testing. What we are looking for is the feature that has catapulted ordinary everyday informal hypothesis-testing into the kind of super-powerful process we find in our research laboratories and field stations. This has much more to do with some supplementary features of formal science that do not really exist in the everyday version.

One such feature is 'modularization' whereby complex phenomena are partitioned into smaller segments that are then dealt with piecemeal. This is the feature that many outside science castigate as reductionism, but which scientists have always insisted is the great strength of their approach. Reductionism is often contrasted with 'holism', and there have been many calls for a more holistic approach to science. Unfortunately the arguments have been much clouded by confusion, so it is important to understand just what is meant by reductionism and the role that it plays in science, as well as what claims are being made on behalf of holism. Part of the problem here is that reductionism can refer to two quite different things: one is theory reduction in the philosophical sense (sometimes referred to as 'nothing but-ism') and the other is methodological reductionism (which is a heuristic device or prescription for how to study the world).

Philosophical reductionism involves the attempt to explain a given phenomenon in terms of theories at some lower level. Thus we might try to explain the phenomena of biology in terms of chemistry, or the phenomena of chemistry in terms of physics. In some respects we do in fact do this: molecular genetics provides us with an account of organisms in terms of their constituent DNA (chemical molecules), while the modern atomic theory of chemistry explains the behaviour of chemical compounds in terms of the physics of the atoms that they are made of. It is in this sense that these disciplines are reductionist: biology is nothing but chemistry, and chemistry is nothing but physics. And sometimes these claims are put into dramatic effect: 'If we could only understand the molecular structure of DNA, we would know all there is to know about organisms' – the claim currently used to justify spending billions of dollars on the Human Genome Project. It's the eternal optimism of science: give me a description of the position of all the atoms in the universe, the great eighteenth-century French mathematician Simon Laplace once declared, and I will use the laws of physics to predict the future state of the universe.

However, this is something very different from methodological reductionism. The essence of the argument here is that nature is far too complex to study as a single entity. The only way to make any progress is to divide it up into more manageable chunks and get to grips with these one by one. Once we have worked out how all the bits work, then we can fit them all back together again and see how they work as a combined system. This is a very different kind of thing from theory reduction, for one of the things we often find on reassembling systems in this way is that they behave in quite different ways to what we might have expected on the basis of the bits and pieces. Systems have 'emergent properties' that are a consequence of their systemic nature. Such effects are particularly common in biology, for example. Strictly speaking they are not a consequence of the fact that the whole is not just the sum of its parts (the claim originally made by Durkheim in defence of the uniqueness of human behaviour), but rather a consequence of our ignorance about how the bits and pieces actually work. It is a bit like examining the way in which a car's radiator system works and then being surprised that, when you connect it up with the engine, it can be used to keep both the engine cool and the passenger compartment warm.

In contrast, the essence of the holistic position seems to be nothing more nor less than the claim that 'the whole is more than just the sum of its parts'. There are at least two ways in which this can be interpreted. One is simply that trying to explain the phenomena of, say, biology in terms of its constituent chemistry ignores the higher-order structural relationships between the parts themselves, or that it ignores other aspects of the living world like the ecological context. It is not entirely clear to me how this differs from methodological reductionism. Basically it just says something that every scientist knows to be true, namely that the world is complicated. In this respect it seems to me that the advocates of the holistic approach seem to have fallen prey to the propaganda generated by new disciplines like molecular biology that have been anxious to cut a dash in the competition for funding. They seem, rather naively, to have swallowed the molecular biologists' view that biology *is* nothing but molecular biology and have failed to realize that there are many other levels at which biologists work.

But holism can also mean something more than this. It is not so much that the world is too complex to be explained by simple disciplines like chemistry, but that there are important components to it that are not physical in the conventional sense and so cannot be studied by science. 'There is more to life than biology' might be an appropriate slogan. The great problem with this view is that it invariably leads us straight into mysticism and religion: belief in supernatural forces often lurks in the background in this version of holism. For this reason alone, we should be deeply suspicious of it, if only because asserting that there are aspects of the world that we cannot understand is tantamount to a surrender of the very capacities that make us human – namely, our intellectual abilities. Worse still, perhaps, such views are reminiscent of the vitalist theories that were so common in biology and elsewhere during the last century which claimed that living matter contains a special force or impulse (an *élan vital*, to use the term coined by the French philosopher Henri Bergson, one of the leading exponents of these ideas) that cannot be studied by the methods of physical science. It took more than half a century to rid biology of the malevolent influence of these ideas and the fact that they still have some considerable vogue outside biology is a depressing reflection on our failure to overcome the superstitions to which we humans seem so prone.

But perhaps the greatest objection is that holism is simply a very pessimistic philosophy. If the real world genuinely is too complicated for us to understand, then there probably isn't too much hope for our future. We will never learn how to control the diseases that strike us down, and we will never be able to save the planet from the fate that two dozen centuries of human mismanagement have left it heir to. And if that's the case, then we might as well go out with a bang rather than a whimper and have a real party on what little remains of the world's resources. In short, holism is a naive recipe for an unmitigated disaster, and we follow it at our peril. Our only real hope for the future lies in the belief that our intellectual abilities are good enough to unravel the complexity of natural processes and allow us to forestall the inevitable fate that awaits us if we don't. It is a race against time, and we cannot afford the luxury of allowing mystical nonsense to distract us from reality, however comforting that nonsense might be.

One other point is worth noting about holism. It seems that people are much less worried about attempts to reduce chemistry to physics than they are about attempts to reduce human behaviour to biology. It is a view that is particularly prevalent in the social sciences, and it owes its origin to Durkheim's claim that human behaviour can only be explained in terms of human behaviour. In part it seems to result from an atavistic fear that by admitting that our behaviour has biological roots we will somehow lose our independence, our free will. It is a strange claim, for it overlooks the fact that most phenomena in the real world can (and have to) be explained at several different levels. Of course, human actions are the product of human beings' thought processes and all that these entail in terms of culture. But they are also the product of brains and, like it or not, brains are biological entities that run on chemical principles. Whatever explanation we can offer for human behaviour in terms of social phenomena can be matched by a set of explanations at the level of neural activity. This is directly analogous to saying that we can offer two different kinds of explanation for how computers work, one in terms of software (the programs) and one in terms of hardware (the silicon chips and the electric currents flying around the rest of the machinery). To say that one is *better* than the other is to miss the point that they are different kinds of explanation. It is a bit like insisting that heads are better than tails on a coin. We can define it that way if we really want to, but it does not exactly give us a full account of what a coin is like if we merely describe its head and refuse to say anything at all about the obverse side.

This separation of explanatory levels has long been a hallmark of biology. As long ago as the 1940s the biologist Julian Huxley pointed out that a biologist could offer at least three very different kinds of explanation for a particular phenomenon. In considering, say, the heart, a biologist might give an account of it (i) in terms of its mechanical properties (its structure, the way it works, the physiological processes that make it beat), (ii) in terms of its function (what it does for you: in immediate terms, pump blood around the body in order to transport oxygen from the lungs to individual cells, but in ultimate terms to enable individuals to maximize their contribution to their species' future gene pool), and (iii) in terms of its ontogeny (the developmental sequence by which specialized cells are produced by the non-specialized sperm and egg cells that came together at conception). To these were later added explanations in terms of phylogeny (the historical sequence of changes that occurred over evolutionary time as one species changed into another, so producing organisms that had hearts from ones that did not).

These four general kinds of explanations (now conventionally referred to as Tinbergen's Four Whys after the Nobel laureate ethologist who first spelled them out in full) provide complementary approaches that depend on very different technical skills. These now constitute separate sub-disciplines within biology. The first involves disciplines like physiology, anatomy, psychology and many of the social sciences; the second involves disciplines like behavioural ecology and population genetics; the third constitutes developmental biology and those areas of psychology interested in nature/nurture questions; and the fourth is made up of disciplines like palaeontology and taxonomy. Each is logically quite independent of the others and none can be meaningfully said to have precedence. A full account of any phenomenon would obviously require all four types of answer. However, because they are logically independent of each other (several alternative mechanisms may subserve the same function, for example, or the same functional outcome may be produced by ontogenetic processes as different as genes and learning), our conclusions about any one type of explanation in no sense commits us to a particular view on any of the others. Nor do we need to be able to answer all of them at once: we can legitimately proceed piecemeal, building up each element of the story independently of the others until eventually we can put them back together again.

While whole organism biologists, in particular, have been used to working with this division of explanatory levels and generally take great pains to avoid confusing them, others have not been so careful. Both developmental biologists and social scientists have commonly confused functional explanations about genetic fitness with ontogenetic explanations about the contributions of nature and nurture (presumably because the term 'gene' happens to occur in both types of explanation and they are not aware that the word refers to different phenomena in the two cases). Similarly, psychologists have commonly confused mechanistic explanations (e.g. about motivations) with functional ones (apparently because they mistakenly interpret the 'teleonomic' nature of evolutionary functional explanations – 'animals always behave in such a way as to maximize their genetic fitness' – as implying purposiveness).

Evolutionary biologists find conflations of this kind very frustrating. To them it is self-obviously the case that each type of explanation is concerned with a different layer of the problem. The ontogenetic processes produce mechanisms with a logic of their own (we cannot infer anything about how these mechanisms work from a knowledge of how they develop from a fertilized egg; all we can say is why they end up looking the way they do); these mechanisms in turn make possible (but do not *determine*) functional consequences that facilitate the propagation of the organism's genes; which, in turn, has implications for the future evolution of that species (but, again, are not solely responsible for its evolutionary history because this also depends on the serendipitous influence of changing environmental conditions – something that is quite independent of the animal's biology).

This last point should remind us that, as biologists have long recognized, we can never predict the future course of evolution: we can never say how the environment will change in the future. All we can ever say is that, all other things being equal and providing conditions remain as they are now, this species will evolve in the following way. But, at the same time, this does not mean that we cannot explain why evolution has taken the course it did. As the philosopher Michael Scriven pointed out more than thirty years ago, evolutionary explanations are *postdictive* rather than predictive: they are designed to provide us with explanations of why things are as they are. They are nonetheless perfectly good scientific explanations because they give explanations in terms of general principles about how the world works.

This highlights a point that most discussions about the evils of reductionism in science invariably ignore, namely the fact that the attempt to reduce one explanatory level to another (the philosopher David Hull has referred to this as 'mechanistic reductionism') simply does not make sense as a research programme. It is perfectly possible to reduce chemistry to quantum physics if you really want to, but doing so yields equations of such unmanageable complexity that we seldom learn anything that we did not already know. Many of the important features of chemistry arise from the emergent properties of chemical systems, from the way in which the physical aspects interact at higher levels of organization. There is no great mystery about this: it is simply a property of complex physical systems. It is a bit like getting excited over the discovery that we can rewrite a computer program that was written in the programming language BASIC in machine code, or better still in terms of the physical changes in the computer's electronic circuitry. Wonderful, but have we learned anything we did not already know about how the program produces its output? A computer scientist would, of course, say 'Yes' because he is interested in how the machine actually works; but the rest of us would say 'No' because we are only interested in the output (what the machine can do for us).

Emergent properties are a particularly important feature of biological systems (if not all real world phenomena). It is these emergent properties that make systems so notoriously difficult to study as interacting wholes. Consequently the only sensible way to proceed is by using a reductionist approach to pull them apart into more or less self-contained bits and then reassemble them bit by bit as we find out how each one works. Refusing to do so is a bit like trying to improve the performance of your car by contemplating it from the other side of the road. There is simply no alternative to opening up the bonnet, dismantling the engine and examining it component by component. Once you have some grasp of what the bits do, you can start to reassemble them in order to determine how they interact with each other. Then, and only then, will you have any real chance of making improvements to the way it works.

On the other hand, of course, there are legitimate questions you *can* answer from the other side of the road without having to lift the bonnet. These include asking what cars are used for (a functional question). No amount of tinkering with the engine will help you answer that question because the engine itself cannot tell us what its purpose in life is. Combustion engines, after all, can be used to run cars, to generate electricity and to operate water pumps, among a dozen other uses. Conversely, rickshaws and cars are both used to transport people from one place to another, even though they bear no resemblance to each other in terms of their mode of propulsion. The function does not tell us about the mechanism, and the mechanism does not necessarily tell us anything about the function. In short, the techniques we use to study the world have to be tailored to the particular kinds of questions we want to ask: there is no single universally applicable methodology in science.

It is for this reason that science is sometimes said to be 'modular'. In principle, if we knew all the ins and outs of how natural phenomena worked, we would be able to reconstruct the whole thing in terms of basic physics, including all the relevant emergent properties of the human mind. But in practice we cannot do it – at least, not yet. The constraint is not imposed by the nature of the phenomenal world itself, but by our own intellectual limitations. Few of us can handle more than two dimensions at once when we try to think about how things work. We have to partition the real world up into more manageable chunks and deal with them in isolation.

It seems that this lesson is lost on both the critics of reductionist science and some of those scientists working on the frontiers of very rapidly moving disciplines like molecular biology. Reductionism in the sense of theory reduction can tell us about the mechanisms that underpin the phenomena we experience. But it cannot answer other important questions about function, ontogeny or evolutionary history. These, as all good biologists know, have to be dealt with separately using more appropriate (albeit equally 'reductionist') methodologies.

6

Unnatural Science

I only ask as my just deserts the liberty I freely allow to all other men, to put forward as true those things which in this whole dark business seem probable until such time as their falsity may be openly proved before all men.

William Harvey: *De Generatione Animalium* (1653)

Whenever one lights upon more exact proofs, then we must be grateful to the discoverer, but for the present we must state what seems plausible.

Aristotle: *De Caelo* (c.330 BC)

Up to this point I have argued that science is a natural phenomenon, common to a wide variety of cultures and organisms. But in doing so I have skirted over the fact that modern science as we practise it in our high-tech laboratories is something more than just everyday knowledge put into practice. In reality, most of the examples that I have discussed have been quite uncontroversial cases of cookbook science (knowing 'that' something is the case). This point has been made very forcefully by Lewis Wolpert in his book *The Unnatural Nature of Science*: the essence of Wolpert's argument is that the findings of science are often counter-intuitive. One obvious example of this is the claim that the earth goes round the sun rather than, as common sense would have it, the sun going round the earth.

We have to be careful not to confuse the message with the messenger here: Wolpert's argument is largely concerned with the *findings* of science, whereas my emphasis has been on the *methods* of science. My point has been that these methods are, at root, simply the natural mechanisms of everyday survival. Nonetheless there are several respects in which the methods of modern science are unnatural. They require us to behave in ways that we might not normally do and this is what often makes science a very hard discipline to follow. Here I shall explore three aspects of science practice that seem to me to be especially at variance with our natural everyday behaviour.

A Philosophy of 'As If'

One of the features of modern science that seems so unnatural concerns the role that theories play. In everyday life, we hold beliefs (sometimes referred to as theories) which are often granted the status of absolute truth. The dogmas of received religion are rarely challenged by the members of the sect that holds them. In contrast, theories in science are merely our current best guesses (even though human frailty often leads us into treating them as absolute truth!). They act like a crutch to help us struggle a little further down the road of exploration in the hope that a more precise theory will be encountered along the way. 'Assume that X is the case . . .' is probably the commonest opening to an argument in scientific papers. For this same reason, scientific papers and even casual coffee-room discussions among scientists are commonly hedged around with an array of ifs, buts and maybes.

An appreciation of this is necessary to understand a common habit among scientists that non-scientists seem to find very perplexing (if not positively disturbing, judging by some of the comments that have been made to me). This is the fact that scientists are prepared to adopt theories that they happily admit are flawed. We now know that the physics of the universe is not Newtonian at all, but something much more complex. Yet scientists proceed to use the principles of Newtonian physics in many different contexts, as though completely oblivious of the fact that Einstein's relativity physics (not to mention modern quantum physics) has radically undermined it. Professional engineers and architects use Newtonian physics, for example, to solve problems of technical design a thousand times a day. You and I use it several dozen times an hour to manoeuvre ourselves from one spot to another, when lifting a fork to the mouth at the dinner table, when throwing balls, deciding when to cross the road in front of an approaching car and in a hundred other trivial daily tasks. Yet, despite the fact that it is fundamentally flawed, we successfully get from A to B, the fork makes it to the mouth, the ball lands roughly where we intend it to and so on. More dramatically, perhaps, its use in the design of the Post Office Tower has not (so far!) resulted in that building's collapse. This is not, I should perhaps add, another example of the much-vaunted 'Five-minute Rule' ('if it stays up for five minutes, it'll probably stay up for good') that medieval architects seem to have used when constructing all those soaring cathedrals. Newtonian physics really does predict well what happens at the scale of the everyday world. It is a reasonable approximation because the features that make it hopelessly inaccurate at the microscopic level (quantum effects) and on the galactic scale (relativity effects) are almost unmeasurable on the scale of the solar system. In effect, scientists operate on a principle of 'As if' – 'Let's proceed *as if* theory X were true (even if we know that it really is not).' If you like, Newtonian physics is simply an emergent property of quantum physics: it's what happens when you scale up and average out all the tiny quantum effects at the sub-microscopic level. The bottom line is that we understand why it turns out to be a reasonable approximation.

Strictly speaking, Newtonian physics must rank as the biggest confidence trick in the history of human learning: it makes all kinds of totally unrealistic assumptions about the existence of perfect vacuums, ideal gases and frictionless processes, none of which ever occur in nature. Every experiment has to be carefully contrived to get it to work, otherwise extraneous variables are likely to produce results that bear no relationship to what the theory predicts. Try dropping a stone and a feather from the same height. Newtonian physics says that their rate of fall is governed only by the effects of gravity and is independent of their respective weights: they should hit the ground together. But, as everyone knows, they don't: the stone will hit the ground before the feather. Aristotle concluded on the basis of everyday experiences of this kind that the rate at which objects fall depends on their weight (or mass, to use the proper physics term). Galileo had to cheat in a famous experiment to show that Aristotle was wrong: instead of dropping a feather and a stone from the Leaning Tower of Pisa, he dropped two cannon-balls of different weights. (Actually, he didn't drop them from the Tower of Pisa at all: he rolled them down an inclined plane. But such is the mythology of science!) The effect is the same, at least in principle, since a light cannon-ball is standing in for a light feather. To show the same effect with a feather, you would have to carry out the experiment in a perfect vacuum, and that was impossible to do in the days before vacuum pumps (and even then isn't all that easy). Instead, we assume that feathers will behave in the same way as small cannon-balls and explain away the feather's odd behaviour as being due to air resistance.

Non-scientists and those who work at the humanities end of the social sciences often find this kind of behaviour especially perplexing. A colleague of mine, who is an evolutionary biologist, recently submitted a paper to an anthropological journal exploring a mathematical model of cultural evolution; a social anthropologist asked to referee the paper got hot under the collar, insisting that so simple a model could not hope to do justice to the extraordinary complexity of human culture, and the paper was rejected. Alas, the anthropologist had missed the entire point of the exercise. Biologists are used to dealing with a biological world of such mind-boggling complexity and variability that human cultural variation pales into insignificance by comparison; they would no more assume that a simple model explains all there is to culture than they would claim that X and Y chromosomes are all there is to sex – if only because sex is determined by X and Y chromosomes in only one insignificant group of organisms (the mammals, the group to which we humans belong). (In fact, there are more ways of determining two sexes in the natural world than there are human marriage systems. I suspect that social anthropologists have been impressed by the variety found in human cultures only because they are blissfully unaware of the range of variation found in the rest of nature.) A mathematical model is not an attempt to provide a complete description of a natural phenomenon; rather, it is an attempt to see how far we can get with a limited set of principles, an attempt to establish the model's boundary conditions.

A classic example of how scientists work in this respect is provided by the story of the discovery of the neutrino. The detailed experimental testing of the so-called Copenhagen model of the atom during the 1920s had uncovered a serious problem: the equation for the decay of the atomic nucleus did not add up to give the same results as those obtained by the experimental physicists. Something was seriously wrong, and it either had to be the whole of theoretical physics or the procedures of the experimentalists. Needless to say, the experimentalists were reluctant to admit that their performance had been anything less than perfect, while the theoreticians defended the accuracy of their calculations with equal vigour. Then in 1930 the young Austrian physicist Wolfgang Pauli proposed an alternative solution that managed to scandalize both sides of the debate at the same time. Pauli suggested that the reason why the equations did not agree with the observations was that a particle had been left out of the theoretical equation. This particle he named the *neutrino* (literally, the tiny neutral particle).

As an hypothesis, this was perhaps a reasonable gambit. What outraged all the physicists was Pauli's apparently cavalier inferences about the particle's properties: these, he declared, were simply whatever was necessary to make the equation for atomic decay balance out with the experimental data. This meant that, amongst other properties, the particle had no electric charge, travelled at near the speed of light and had no mass. To have a chargeless particle was by no means unusual, but a particle with zero mass that travelled at the speed of light was a contradiction in terms: under the tenets of the Theory of Relativity, mass is proportional to speed, and anything travelling at the speed of light would have infinite mass (which is why nothing can travel faster than the speed of light). What made the situation worse was the fact that this particular combination of properties made it virtually impossible to detect the particle by any conventional means. Pauli's hypothesis had all the worst hallmarks of the kind of *ad hoc* gap-filling that scientists so dislike. The reason scientists abhor such behaviour is that it does not tell you anything new; it's simply another way of describing the world as you already know it. Pauli appeared to be saying, in effect, that his new theory was simply the bits that were missing from the Copenhagen model, despite the fact that he had no independent theoretical grounds for being able to make that claim.

However, it soon became apparent to the physics community that they had little choice but to accept Pauli's suggestion. The problem was that if they rejected it, they would also be forced to reject an even more fundamental principle in physics, namely the principle of the conservation of forces. Since that would make even Newtonian physics impossible to do, the physicists set about building detectors that would be able to register any neutrinos that happened to pass by. Unfortunately the calculations showed that, thanks to a combination of their enormous speed and negligible mass and the fact that modern atomic theory tells us that matter is mostly empty space, neutrinos would pass through most forms of matter without being noticed. Even if they existed in enormous numbers, they would be extremely difficult to detect unless they first passed through so much 'solid' matter that their speed was slowed down to more conventional levels. By dint of placing enormous vats of pure water at the base of the deepest mines in existence, physicists were eventually able to detect three neutrinos some twenty-five years after Pauli's original launching of the idea. Since then, more neutrinos have been detected by similar instruments. In one recent case, an unexpected burst of neutrinos even alerted us to a supernova explosion elsewhere in the universe, which was then detected a few days later by more conventional telescopes.

Pauli had been right after all: neutrinos did exist and they did have the properties that he suggested. Only on the question of the neutrino's mass was Pauli wrong: the neutrino does turn out to have a mass, but it is so small that physicists have so far only been able to place an upper limit on it. More interestingly, perhaps, neutrinos are probably much more common than physicists originally supposed, for they are constantly being created in the nuclear reactions of stars. In fact, some calculations suggest that they may be so common that the combined weight of all the free neutrinos in the universe may actually exceed the total weight of visible matter. So much for an *ad hoc* stopgap. Yet scientists would have been seriously misled had they adhered too strictly to naive Popperian views of how science should be done. Instead, success emerged from their willingness to suspend belief and proceed as if the neutrino hypothesis were true (even though many of them thought it was probably wrong). But notice that they did not behave in this way out of pure whimsy. In no sense was the decision to go along with Pauli an arbitrary or capricious one. They did so with some misgivings, and only because it seemed the least bad option given the structure of the rest of physics as it was then understood.

That scientists behave in this way is largely justified by experience. I remember one amusing instance of this at an international conference held in 1982 to celebrate the centenary of Darwin's death. The evolutionary biologist John Maynard Smith had just given a lecture on the application of mathematical game theory (a branch of economics) to the study of animal behaviour. When the chairman called for questions from the audience, a well-known neurobiologist (who happened, for rather obscure political reasons, to disapprove rather strongly of this kind of approach to human behaviour) asked whether it was sensible to make so many assumptions about the genetic basis of behaviour and the validity of economists' mathematics when discussing human behaviour. Maynard Smith squinted through his glasses for a moment before responding. 'During the war,' he said, 'I was an engineer working in aircraft design. We made all sorts of assumptions and took all kinds of shortcuts on the physics of flight, but the aircraft we built still flew.'

In this respect, it seems to me, trained scientists do differ from both ordinary people and academics in many of the non-scientific disciplines. In everyday life we want certainty and we invariably view our theories as Absolute Truth. We need solutions *right now* for the problems of poverty and environmental degradation, or to cure particular diseases. We cannot afford to wait until the year 2500 AD for the answer, since failure to act now may find us bedfellows with the dinosaur. When a child is sick, we want a cure now, not in half a century's time. But scientists operate on a different time-scale. Their concern is not to try to solve all the problems today (although occasionally being able to solve one problem in a lifetime comes as a bonus); rather, their concern is to understand how and why the world is as it is, and if this takes five hundred years of collective effort by a thousand individuals scattered in laboratories all around the world, only a handful of whom ever actually get to meet, then so be it. Success in science comes only from a long slow methodical working through of all the ins and outs of a very complex phenomenon, checking and double-checking everyone else's calculations, because only by patience and careful testing will we avoid mistakes. As the history of science all too often reminds us: nothing is gained by over-hastiness.

Strong Inference

A second feature of modern science that makes it a demanding intellectual task is the rigour with which ideas are worked out and tested. Our ability to test hypotheses ultimately depends on how precisely these hypotheses can be formulated. The more precise the hypothesis, the more specific the prediction it makes, and hence the more powerful the test we can perform simply because it is then harder for empirical results to appear to confirm the prediction if the hypothesis is not in fact true. The significance of this for the progress of science was first recognized by J. R. Platt in an article published in the prestigious American journal *Science* in 1964. Platt was mainly interested in the reasons why molecular biology had made such spectacular progress. In 1940 the discipline did not exist; by 1960 it was set to become a major force in biology; and by 1970 it had taken over as the dominant component both financially in terms of the proportion of national science budgets allocated to it and numerically in terms of staffing. Its success has been extraordinary, even by the standards of science.

Platt asked why this had happened in so short a space of time in this particular discipline when other branches of biology, like taxonomy (the sub-discipline that deals with the classification of species), had remained more or less stagnant during the same period. His answer was that molecular biology had been able to capitalize on the powerful theories of chemistry. This meant that molecular biologists (unlike their whole-organism counterparts in ecology, behaviour and taxonomy) could generate very precise predictions to test. Rather than simply describing what happened (as in traditional inductive science) or formulating rather vague predictions (as in the early stages of explanatory science), molecular biologists had been able to say: 'If the experiment does not produce *exactly* result X, then the hypothesis must be wrong.' This meant that hypotheses were put through a much more rigorous form of testing. As a direct result, molecular biologists were able to weed out the incorrect hypotheses more quickly. Rather than being misled by results compatible with several mutually contradictory hypotheses, they were able to scrap whole armies of alternatives at a stroke and get on to the next question with a great deal more certainty about the conclusions they had drawn. Platt termed this *strong inference*.

The point is that a great many factors may be generally compatible with a rather weak hypothesis that simply says 'X is bigger than Y', but many fewer hypotheses will be compatible with the prediction that 'X is exactly five times bigger than Y'. Recall our Greek peasant farmer back in the time of Aristotle. Suppose he believes that his barley grows better if planted at the onset of the zodiacal sign of Pisces in mid-February. His hypothesis, in effect, is that barley crops planted in Pisces are larger than those planted in Aries (beginning in mid-March). And since, in general, this turns out to be the case, he legitimately concludes that his hypothesis is right. But it only appears to be right because the spring rains that occur in March have mostly soaked away by April, and it is the amount of rain in the month of germination that is really determining plant growth. Had his hypothesis been that Pisces crops are exactly five times larger than Aries crops, then the story would have been rather different. For he would have noticed that fields planted in February sometimes produce twice as much as fields planted in March and in other years eight times as much, but that they occasionally produce only half as much. In reality it all depends on both the timing and the quantity of rain in any given year. Sometimes the spring rains are late and occur in April; sometimes they are rather weak; and sometimes they are unusually heavy. A more precisely formulated hypothesis would have allowed him to make predictions that were compatible with fewer alternative hypotheses. He would have been forced to reject his signs of the zodiac hypothesis and begin the search for a better alternative explanation that much sooner.

Other sub-disciplines within biology like ethology (the study of animal behaviour) have had a similar lifespan to molecular biology. (Ethology first appeared in the 1930s, barely a decade before the publication in 1944 of the paper by Avery and his collaborators that is generally credited with being the starting-point for molecular biology.) Yet, by comparison, ethologists were still working in the Dark Ages. It was to be another two decades or so before the same dramatic process began to occur in ethology, but by then molecular biology had swept the board.

Nonetheless when ethology's own revolution came about in the late 1970s, it did so for much the same reason. What is now generally referred to as the 'sociobiological revolution' was made possible when ethologists finally realized how they could apply the very precise mathematical techniques of population genetics to something as nebulous as behaviour. The groundwork had in fact been laid a decade earlier through the work of the theoretical biologist Bill Hamilton. In 1964 he published a pair of technically rather formidable papers that provided a Darwinian solution to the problem of how altruistic behaviour could evolve. Darwin himself had recognized the problem a century earlier. If an organism dies while helping to save someone else (for example, by giving an alarm call on sighting an approaching predator, so attracting the predator's attention to itself while allowing everyone else to take evasive action undetected), the genes that compel it to behave in this way will very quickly be wiped out from the population by natural selection; yet altruistic behaviour of this kind quite obviously continues to be perpetuated within many animal species.

What Hamilton was able to show was that, so long as its altruistic actions enhanced the rate at which relatives who possessed copies of the same gene were able to reproduce, a gene for altruism would be successfully propagated in future generations even if the individual behaving in an altruistic fashion never reproduced. From the gene's point of view, it doesn't matter who actually does the reproducing so long as copies of it find their way into the next generation. The principle (now known as Hamilton's Rule) is very simple, but proving this mathematically turned out to be rather a nightmare. It took more than a decade for the implications of these rather impenetrable papers to filter through into the working consciousness of the ethologists. But once they did, this biological sub-discipline took off too. It has not achieved the dominance of molecular biology, but this is hardly surprising: despite its importance for areas like animal welfare, ethology's lack of immediate biomedical and pharmaceutical relevance has inevitably meant that its funding has remained meagre by comparison with that given to disciplines that hold out the promise of medical cures and large commercial profits.

The history of astronomy offers us another, perhaps even more dramatic, example of the same phenomenon. Astronomy is, without question, the oldest of the sciences. The Chinese, the Babylonians and the Egyptians were all busily engaged in star-gazing at the very dawn of written history 6000 or so years ago, and had no doubt been doing so for millennia before then. Indeed, the French archaeologist Alexander Marshack has interpreted the series of notches carved on ivory and bone pieces from various European cave sites as evidence of crude calendars recording the passing of the lunar months as early as thirty thousand years ago. The so-called Blanchard bone (a small sliver of bone barely four inches long discovered in 1911 at the Blanchard rock shelter near the French village of Les Eyzies) is pitted with a series of sixty-nine circular and semicircular markings that Marshack argues is a record of the waxing and waning of the moon over a two-month period. If Marshack is right (and not everyone agrees with him), then monitoring the changing patterns of the night sky may be one of the oldest activities in the history of human thought.

But only with the rise of physics in the seventeenth century did astronomy begin to aspire to being more than a purely descriptive science. Even so, it remained small-scale, confined to the immediate problems of the solar system and the mathematics of planetary orbits. It was to be a further three centuries before it was to 'take off'. But when it did, it did so with a vengeance. Within a century, astronomers would be asking questions about the composition of distant stars and the evolution of the universe; they would be testing models of the origins of the cosmos fifteen billion years ago and trying to decide whether the universe would continue to expand for ever or collapse on itself in a catastrophic gravitational implosion.

It has been possible to do this only because astronomers were able to capitalize on the mathematically rigorous theories of physics. In thinking about such remote events as the Big Bang at the very start of our universe, they have been able to formulate very precise models for the initial conditions, and then determine what their observable consequences would be fifteen billion years later by allowing the laws of physics to work themselves out from these different starting-points. Then, by matching the observed state of the universe as we see it now to the predicted outcomes generated by these models, they have been able to infer a great deal about the initial conditions during and soon after the Big Bang. Simon Laplace would have been all smiles.

Strong inference has played a particularly important role in almost every scientific field that has undergone such a phase of rapid development. It acts like a catalyst, speeding up the rate at which progress can be made because you can be more certain about the conclusions you draw. Precisely formulated hypotheses are compatible with a very much narrower range of empirical results than more loosely formulated ones. That makes it much more difficult to confirm an hypothesis by accident or because of the influence of extraneous confounding variables. Mathematics and the mathematical theories of physics that had been developed by Newton and the classical physicists between the last quarter of the seventeenth century and the middle of the nineteenth were an essential ingredient in all this. Directly or indirectly, each of the fields that went through a sudden revolution of this kind did so because it was able to exploit either the theories of physics directly or the mathematical tools developed to solve the problems of physics.

Evolutionary biology and the study of animal behaviour, for example, benefited from being able to apply the mathematics developed by economists for finding optimal solutions to investment and pricing problems. Likewise, biologists have made extensive use over the past two decades of mathematical game theory: this was originally developed by economists to solve military operations problems, but was later applied to pricing wars and labour-relations strategies. Evolutionary biologists have used the mathematical techniques of game theory to understand patterns of genetic evolution as well as the evolutionary aspects of animal behaviour. It has been used to explain, for example, why animals rarely fight to the death but more often settle disputes using conventions such as who can roar the loudest.

But the biologists did not simply import economics into the study of behaviour: they did not borrow the economist's entire world-view and *become* economists. Rather, they borrowed only the economist's mathematical tools for finding the maxima and minima of functions or the optimal solutions to decision problems, just as the economists had, in their turn, borrowed these techniques from the physicists who had invented them to solve the problems of fluid dynamics and light propagation. At each borrowing, the mathematical techniques were developed and elaborated in new ways to deal with different kinds of problems. One example of this is the fact that economists are now borrowing back from ethologists some of the mathematical approaches they originally invented, but now reconstituted to answer new problems that relate to how organisms deal with risk (a problem that the ethologists encountered in studying animals' foraging behaviour).

Physics itself had, of course, gone through its own dramatic change of fortunes during the Newtonian era. In this case the physicists had not borrowed a powerful set of theories from another discipline; instead, they had actually created the theories they needed for themselves. But they had been able to do this only because of two special conditions.

One is the fact that the kinds of phenomena they were dealing with at the time were fairly simple and very regular in their behaviour. Remember that what set the ball rolling in the seventeenth century was an interest in the mechanics of planetary orbits. Fortuitously, these had many similarities to the behaviour of pendulums and other simple physical systems that could easily be studied in the laboratory (or for that matter the naves of very large churches in the case of Foucault's celebrated pendulum). These processes are also rather regular in their behaviour: they are not determined by a particularly complex set of interacting variables and therefore repeated experiments tend to produce fairly consistent results. Consequently they proved to be relatively easy problems to analyse.

The second special condition that made the rise of physics at the end of the seventeenth century possible was the fact that the mathematical tools needed to describe these phenomena had been developed by a long series of mathematical geniuses, beginning with the ancient mathematicians of Egypt, Babylon and India, and the Greek and Arab worlds. Men like Euclid, Eudoxus and Archimedes among the Greeks, and al-Khwarizmi, Thabit ibn Qurra, Omar Khayyam and Jamshid al-Kashi among the medieval Arabs, had provided the basis from which a whole host of mathematical techniques were to be developed over the ensuing centuries by a group of mathematical wizards (among them Euler, Pascal, Descartes, Leibnitz and, of course, Newton) that flowered in western Europe during the seventeenth and eighteenth centuries. Archimedes' discoveries were particularly important in this respect because they laid the foundations for the theories of hydrostatics and mechanics by showing that many of these problems could be described and solved mathematically. Not only did Archimedes all but invent the integral calculus nearly two millennia before Newton and Leibnitz, but he also came within a whisker of pre-empting them on differential calculus too. It would not be too much of an injustice to say that, without the benefit of these two key mathematical tools, Newton might have been nothing more than an interesting footnote in the history book of science.

The Logical Mind

Strong inference is very much bound up with the processes of logical argument. In developing predictions to be tested, we normally make a series of inferences from a set of premises (or assumptions) based on some model or hypothesis about the way the world is. This requires that the logical processes of inference be rigorously accurate, for any sloppiness in the chain of inference will lead to inaccurate (even completely incorrect) predictions. More than anything else, it is perhaps the logical rigour with which scientific arguments are pursued that the man-in-the-street finds hardest to cope with. More often than not in the context of everyday discussions, the arguments we use are sloppily constructed. Conclusions do not follow from the premises, *non sequiturs* (in which statements are wholly unrelated to anything that has been said before) abound. The precision and powers of logic that a scientist applies (or at least tries to apply) often strike the layman as inhuman: the best scientists seem more like robotic calculating machines than real living people. Here is not the evil fiendishness of Mary Shelley's Dr Frankenstein, but rather the cold emotionless logic of Star Trek's Mr Spock, devoid of the human qualities of warmth and honest frailty.

That people find it difficult to think through the logical structure of arguments was recognized by Aristotle as long ago as the fourth century BC. To try to overcome this, he codified the rules of inference as a set of abstract syllogisms. Each syllogism consists of three propositions (two premises or assumptions and the conclusion that can be drawn from them). A particularly simple example might be:

Premise 1	All professors are human	(All A are B)
Premise 2	All humans are mortal	(All B are C)
Conclusion	Hence all professors are mortal	(∴ All A are C)

This is the very same syllogism that the rat learned to solve in the Holland and Straub experiment I described in Chapter 4: 'Light means food; food means feeling sick; therefore, light means feeling sick – so don't approach the hopper.'

On the whole, these rules have stood the test of time and remain succinct guides to both sound inference and common logical fallacies for most everyday purposes. For the more sophisticated purposes of serious argumentation, however, they have turned out to be less satisfactory, in part because they force us to construct our arguments in a rather stylized form. New forms of logic have been developed during the past two centuries that are both more flexible and more powerful as ways of representing and analysing the structure of arguments. Nonetheless, Aristotle's original observation remains unchanged: humans are not very good at constructing rational arguments and need a helping hand.

That we find it difficult to think logically has been demonstrated repeatedly by psychologists. Some years ago the British psychologist Peter Wason devised a test of logical thinking that was designed to mimic the hypothetico-deductive mode on which science is based. The test (now known as the Wason Selection Task) was intended to assess the philosopher Karl Popper's claim that scientific method is in some sense natural. It was a very simple test: the subject is presented with four cards that are labelled A, D, 3 and 6. It is explained that cards with a vowel on one side always have an even number on the reverse side. (This, if you like, is a kind of generalization, and so deemed to be equivalent to a theory in science.) Which card(s) would you need to turn over to check whether this generalization is true? The answer is the A and the 3: there must be an even number on the back of the A and there cannot be a vowel on the back of the 3, but the rule specifies nothing about consonant or even number cards, so you can have anything you like on the reverse faces of the D and 6 cards. According to Aristotelian logic, a proposition of the form *If P then Q* ('If it's a vowel, then there must be an even number on the reverse') can only be falsified by showing that *Q* is false when *P* is true (i.e. by showing that *P* and *not-Q* are simultaneously true – that a vowel has an odd number (a not-even-number) on its reverse). Typically, less than a quarter of those tested on this task get it right, a result that has held up under repeated tests with a wide variety of subjects on at least two continents over the last thirty years. Most people choose either the A card on its own or the A and the 6 cards. Very few choose to pick up the 3.

People's ability to think rationally in this way has been a subject of some considerable interest in the last few decades. There are practical reasons for this interest, not least the fact that assumptions about rational behaviour underpin our modern theories of economics. Economists assume that people will choose to do things on the basis of a fairly straightforward assessment of the costs and benefits. So if you want to discourage people from doing something, you need only to raise the costs or lower the benefits. Sometimes it works (offering incentives for investing surplus income in savings schemes seems to be attractive), but sometimes it doesn't (increasing the fine for parking in no-parking areas doesn't seem to work, unless the punishment is draconian – for example wheel-clamping – and the probability of being caught a near certainty).

People seem to be especially bad at discounting probabilistic effects. They invariably worry more about being killed while travelling on a plane than when travelling by car. Their sense of proportion, it seems, is overwhelmed by the fact that when a jumbo jet crashes 350 people are killed, even though such crashes occur so infrequently compared to automobile crashes that the risk of dying as a passenger on a commercial air flight is trivially small compared to the risk of dying as a passenger in a car. The rational thing to do is to avoid travelling in cars, yet it's always planes that people avoid if they avoid anything.

The American psychologists Amos Tversky and Daniel Kahneman found that people often have surprising difficulty interpreting probabilities. Although we live in a probabilistic world (nothing, aside perhaps from the proverbial death and taxes, ever happens with absolute certainty), we tend to think predominantly in terms of simple dichotomies. Either something is the case or it isn't. This is particularly so when it comes to trying to evaluate the implications of events when making decisions on what to do in the future.

To examine the way people make decisions, Tversky and Kahneman presented subjects with verbal problems. One group of subjects would be told to imagine that the government is preparing for an outbreak of an infectious disease such as Asian flu that is expected to kill 600 people. They have to decide between two alternative health programmes that can be expected to have different consequences for the population. If programme A is implemented, then exactly 200 people will be saved, but if programme B is adopted there is a probability of one third that all 600 will be saved and a probability of two-thirds that no one will be saved. 72 per cent of those presented with this problem chose programme A. However, when a second group of subjects was told that programme A would result in 400 people dying whereas programme B had a probability of one third that no one would die and two-thirds that all 600 would die, 78 per cent chose programme B. Tversky and Kahneman obtained similar results from widely different groups of individuals, including students, university lecturers and medical doctors.

These results are interesting in two different respects. First, people do not seem to respond in a rational way to statistical problems that, in effect, involve making bets. The two programmes in each case are identical in terms of their long-term outcome as probability problems: on average, both will result in about 200 people being saved and 400 dying. If people behaved rationally, they should choose the two programes with equal frequency because there is no way to decide between them. They clearly do not do so, however, but prefer instead to choose what looks like greater certainty. Secondly, the two problems are identical, except for the fact that one is phrased in terms of survivors and the other in terms of deaths. Even if they prefer certain strategies to risky ones, people should respond in exactly the same way on the two problems. But they don't. People presented with the first problem respond in a way that is risk-averse (they avoid what they see as the risky decision that might result in more deaths) whereas people presented with the second version respond in a risk-prone way (the certain deaths of 400 people seem to be less acceptable than the odds-on probability that all 600 will in fact die). The way a problem is phrased thus seems to tilt our response in one direction or the other. It's as if we look for the best option in each case and go for that, even though the probability of that option actually occurring is quite small. A small chance that 600 might be saved (even though there is a strong likelihood of all 600 dying) seems to be preferred to the certainty that only 200 will be saved. It doesn't make any statistical sense. Yet, these results are extremely robust.

Tversky and Kahneman have examined a number of other reasoning processes in humans. Another classic problem in probability theory concerns conjunctions (the likelihood that two or more things are true of the same individual). The likelihood of a conjunction being true cannot be greater than the likelihood that each of the constituent elements is true. For example, the likelihood that Mary is both a school teacher and a tennis player cannot be greater than the likelihood that she is a school teacher (given the proportion of Marys that are school teachers) or the likelihood that she plays tennis (given the proportion of Marys that play tennis). The laws of probability state quite simply that the likelihood of both being true of Mary at the same time is precisely the product of the two individual probabilities. If the probability of Mary being a school teacher is 0.2 (meaning that 20 per cent of all people named Mary are school teachers) and the probability of her playing tennis is 0.3 (meaning 30 per cent of all Marys play tennis), then the probability that Mary is *both* a school teacher *and* a tennis player is exactly 0.2 × 0.3 = 0.06, which is obviously a lot smaller than either of the two individual traits being true.

Tversky and Kahneman found that people invariably get problems of this kind wrong: typically 80–85 per cent of people will rate the statement 'Mary is both a teacher and a tennis player' as being more likely to be true than the statement 'Mary is a teacher.' It seems as though they are particularly likely to be misled if one or other of the traits is known to be especially characteristic of the type of individual concerned. The fact that some teachers are tennis players misleads us into thinking that both traits are likely to be true.

Like the Wason Selection Task, this is a hard one to see through. The conjoint event does *look* more likely and you need to think very carefully to be convinced that the mathematicians are right. Here's a similar, but much easier, task that Tversky and Kahneman tried on a group of students at the University of British Columbia in Vancouver. When they were given sixty seconds to list all the seven-letter words they could think of that ended in *-ing* and then another sixty seconds to list all those with endings in the form -*.n.*, the students produced an average of 6.4 words in the first category and just 2.9 in the second. Yet all the words in the first category are included in the second (along with a great many other words, of course): the second category *must* be bigger than the first. It seemed that they simply didn't realize that this must be true. And it wasn't just a problem of memory and recognizing words. When another group was asked to estimate the number of words of the two types in an average 2000-word block of text from a novel, they averaged 13.4 *-ing* words and 4.7 -*.n.* words. More worrying, perhaps, is the fact that when Tversky and Kahneman tried a similar test on hospital doctors (with disease symptoms as the variables), even the doctors made the same mistake!

It seems that people resort to comparing scenarios when evaluating different situations and base their decision on the scenario which seems the more probable. It's back to mental models again. A scenario that has a familiar ring to it *seems* more likely than one that is less familiar. Humans were not designed by evolution to evaluate probabilities carefully, because there was no pressing need to do so. So long as they get it right most of the time, that is good enough. And getting it right most of the time seems to involve learning a few key rules of thumb: 'Go for the more familiar situation because it is more likely to be right!' In the informal situations of everyday life, logical errors are not too serious because we usually have time to recognize and rectify any major disasters before they actually occur. It's only in the precision world of science that we come seriously unstuck.

Science's success hinges on a very rigorous application of the principles of logical deduction and the meticulous testing of hypotheses. Mathematics has played a particularly important role in both respects because it allows us to express very complex ideas in ways that, first, force us to state quite explicitly the assumptions we are making and, second, allow us to carry out calculations that are beyond our immediate powers of intuition.

Although these processes derive from common sense, the rigour with which they are applied in science is genuinely unnatural. We find it hard to sustain that rigour because we do not naturally think in these ways. For scientists as much as laymen, the restrictions of the scientific method become irksome: we want to leap ahead, to jump to conclusions, to act impetuously on an exciting new idea and to bask in the glory of everyone's recognition of our achievement. Part of the reason for this, as I suggested in Chapter 5, is due to the fact that such detailed analysis is hopelessly inefficient in the context of everyday survival. Our minds have been honed by evolution to identify rules of thumb that suffice for everyday purposes as quickly as possible. But there is another factor that seems to be important in the context of the more conventional sciences, and this is the fact that our minds seem to be predisposed to deal with social matters rather than the nature of the physical world. It is to this last but nonetheless crucial element in our story that we must now turn.

7
The Social Brain

I suspect that you wonder whether I realise how hard it is for you to be sure that you understand whether I mean to be saying that you can recognise that I can believe you to want me to explain that most of us can keep track of only about five or six orders [of intentionality].

Dan Dennett: *Behavioral and Brain Sciences* (1984)

We humans seem to have a particular fascination for the complexities of the social world. By comparison, our interest in the physical world that surrounds us seems to be desultory, at best. One obvious piece of evidence for this is the fact that biographies and fictional stories (especially romances) dominate the book publishers' lists, far outstripping sales of non-fiction (where everything from musical appreciation to travel guides to popular science gets lumped together).

Indeed, it is so much easier for us to think in social terms that we commonly ascribe motivations to the physical world. In traditional societies, inanimate objects like volcanoes may be attributed personalities: their eruptions are said to reflect their anger with us. We are no less prone to this odd habit in technologically more advanced societies, it seems, for our language is riddled with anthropomorphisms of just this kind: we speak of the wind as being 'vicious', of the sea as 'raging', of a dark and lowering sky as 'angry'. 'I tax not you, you elements, with unkindness,' wails the aged King Lear in his moment of torment on the heath, and the evangelist John has Jesus remark in the course of his conversation with Nicodemus that 'the wind bloweth where it listeth.' Nor do such allusions merely reflect ancient modes of thought: the poet W. B. Yeats alludes to the 'murderous innocence of the sea' in his poem 'A Prayer for my Daughter', while A. E. Housman observes in his poem 'On Wenlock Edge' ''Tis the old wind in the old anger'.

I suggested in Chapter 4 that children behave as natural scientists, formulating and testing hypotheses about how the world works. But recall that the focus of their attention is, in fact, on animate objects. Children very quickly begin to see certain kinds of objects as having interactive properties: they play with you and may even be persuaded to give you things you cannot reach. Their interest in the inanimate world is very much less intense, especially once they realize that it does not have a mind of its own. They will gaze and gurgle at a human face for minutes at a time where a toy will barely hold their attention for a few seconds. We humans are social beings and, from beginning to end, our lives are coloured and dominated by the need to interact with each other. Is it possible that this apparent sensitivity to other social beings accounts for the difficulty we have in dealing with the non-social world, and so explains why we find science so difficult to do?

A Window on Our Minds

The nature of mind has long held a special fascination for us. Philosophers have puzzled about it since time immemorial, some arguing that the mind (or 'soul') is separate from the body, others that it is 'nothing but' the way the brain's activity appears to us. In recent years interest in the mind and its contents has been the subject of renewed interest thanks to the suggestion that the human mind is specially adapted to handling the complexities of the social world. Evidence that this may be so comes from an elegant series of experiments by the American psychologist Leda Cosmides.

Cosmides used the Wason Selection Task that we met in Chapter 6. Now you might argue that the reason why so many people get it wrong is that the task is too abstract for the average human to cope with. We are just not used to thinking in terms of numbers and letters in this way. Cosmides wondered whether subjects would do better with more concrete examples that concerned everyday experiences of the world. She tried tasks relating to the use of urban transportation ('If a person goes to Boston, then he always takes the subway,' with the task presented as four individuals about whom it is known that one is in Boston, one in Cambridge, one came by car and one by subway) as well as tasks relating to foods ('If a person eats hot chilli peppers, then he will drink a cold beer'). Irrespective of the nature of the task, subjects perform only slightly better than with the abstract Wason version. About half the subjects now get it right.

However, when the task is presented as a purely social problem ('If a person is drinking beer, then he must be at least twenty-one years old' or, less familiarly for Western subjects, 'Eating cassava roots is only permitted for married men'), most people get it right. It seems somehow obvious that, when presented with four individuals who are, respectively, drinking beer, drinking Coca-Cola, sixteen years old and twenty-five years old, we need to check on the beer-drinker's age and the sixteen-year-old's drink. How old the Coca-Cola-drinker is or what the twenty-five-year-old happens to be drinking are quite irrelevant to a social rule that stipulates that only those over twenty-one are allowed to consume alcohol. The problem is now the very familiar one of detecting infringements of social rules.

Cosmides and her colleague John Tooby have argued that the human mind is specially adapted for detecting social cheats – that is, people who do not fulfil their social obligations or abide by the rules that society has evolved to enable it to function smoothly. Their argument is based on the assumption that when the human brain evolved into its present form some time in the late Pleistocene 200,000 years ago, it evolved mechanisms needed for handling the really important problems of everyday life current at that time and these problems were the social ones of maintaining the cohesion of our social groups. The crucial problem that our ancestors faced, they suggest, is that of the free-rider, the person that cheats on social conventions by refusing to repay debts or abide by conventional rules. It is the person who repeatedly borrows coffee from you and never pays it back, who never buys a round of drinks when out with a group, who begs favours without returning them; it's the one who parks in the no-parking area so blocking the traffic for everyone else, who grazes more cattle on the common pastureland than he is entitled to by the conventions of the village, who avoids paying quite as much tax as he ought to by exploiting some ingenious loophole in the law, who steals your property rather than working to earn the money to buy his own. It's what Garrett Hardin termed the 'tragedy of the commons': the fact that it invariably pays everyone in the short term to cheat just a little on the system even though in the long term they would all do better to co-operate with each other by adhering to society's rules.

Cosmides and Tooby's argument builds on earlier work by the evolutionary biologist Robert Trivers who suggested in 1971 that what he termed 'reciprocal altruism' was an evolutionarily viable strategy in a Darwinian world. It's the I'll-scratch-your-back-if-you-scratch-mine principle. The only problem is that it is rather susceptible to cheats who enjoy their backs being scratched, but then refuse to repay the debt. Despite this, reciprocal altruism can, nonetheless, be a successful strategy providing the mutual advantages are beneficial and there is some kind of mechanism for detecting and punishing cheats. Generally speaking, the critical requirements are that you have the opportunity to meet repeatedly with the same prospective partner and that you have the ability to remember who did what the last time you met.

In such contexts, the best strategy turns out to be a very simple one. In the technical literature of evolutionary biology it's called 'Tit-for-Tat' (or TFT for short). It involves behaving co-operatively the first time you meet someone and thereafter simply behaving in exactly the same way as your opponent did on the previous occasion. If your opponent refuses to scratch your back after you have scratched his, then next time you meet you should also refuse; if he scratched your back last time, then you should scratch his again next time. It's an extraordinarily simple rule of behaviour, but it works better than any other, as the biologists Robert Axelrod and Bill Hamilton convincingly demonstrated when they held an open competition for solutions to this problem. Axelrod and Hamilton invited game theorists and other mathematicians to submit strategies (or game rules) for playing this game and then ran the rules against each other in a computerized tournament. Each strategy was played against every other strategy many thousands of times, and the strategy that consistently won was the simplest of them all, TFT.

The problem for TFT is that it depends on your repeatedly playing against the same individual. In a one-off game, cheating always pays. If you can arrange your life as a series of single encounters, then cheating on a social convention is a very profitable strategy. One way for cheats to engineer this is to be constantly on the move, always one step ahead of the individuals that have rumbled their cheating ways. The Swedish biologists Magnus Enquist and Otto Leimar were able to show mathematically that free-riding becomes an increasingly successful strategy as both the size and the patchiness of the population increase. The more the population is broken up into small discrete units that communicate rarely with each other, the more easily free-riders can survive because they can move on to another group each time their cheating habits are discovered. These are obviously features that are particularly characteristic of human societies, both traditional and modern.

So, argue Cosmides and Tooby, humans have evolved minds that are especially sensitive to those who cheat on social rules. Humans live in a particularly social world: the social world has, if you like, been our primary evolutionary adaptation, our way of solving the Darwinian problems of survival and successful reproduction. Our survival in the ecological world depends crucially on the success with which we co-operate with each other, for in the natural world we are beset not only by the conventional array of predators but also by marauding bands of humans searching out opportunities to prey on the unsuspecting. Nonetheless, so long as most people co-operate and abide by the rules, it will always pay some people to exploit them and become free-riders.

Co-operation occurs in a great many contexts in human societies. In most hunter-gatherer societies, for example, the produce from a hunt is always shared around the group. The hunter brings his kill back to camp, where it is divided equally between all the families present. Hunting is a risky business, in which successful kills occur at infrequent intervals and meat-sharing probably serves to minimize the risk that any one family will have to go without for any length of time. It seems to be a classic example of long-term reciprocal altruism: I'll let you share in my good fortune now, providing you let me share in yours later when I've been unsuccessful. But hunters always have the option of cheating because they can eat some or all of their kill while still out in the forest, and sometimes they do just that. The trouble is that if too many hunters behaved in this way, the benefits of sharing meat would be lost. So the practice is frowned on as morally reprehensible, and offenders risk finding themselves socially ostracized. All human societies use smear campaigns and snide remarks to enforce the social graces. And we are ever-watchful to see that the rules are being adhered to (at least by everyone else, even if not always by ourselves).

How Monkeys See the World

The claim that the human mind is especially adapted to handling social problems was spun off another more general principle known as the Machiavellian Intelligence hypothesis. Although first mooted quite independently some years earlier by a number of ethologists (including Michael Chance, Alison Jolly and Nick Humphrey), the hypothesis was formally proposed in the late 1980s by two British psychologists, Dick Byrne and Andrew Whiten, as an explanation for the fact that monkeys and apes (the primates) have much larger brains (relative to their body size) than any other group of animals. The fact that primates have such large brains had been known for a long time, but no one had been able to suggest a convincing explanation for it. The Machiavellian Intelligence hypothesis attributed primates' larger-than-average brains to the fact that they live in much more complex social groups than all other animals. The social world of the average monkey or ape differs in two key respects from those of other mammals and birds: one is the way they form alliances against each other and the second is the extent to which they use tactical deception (hence the 'Machiavellian' component: Nicolo Machiavelli was the Italian Renaissance philosopher who wrote a book called *The Prince* providing advice on the kinds of devious political strategies that a ruler should employ in order to outwit his opponents).

Primate coalitions are much more complex than the kinds of coalitions found in other groups of animals. Whereas animals like swans and lions may call on each other for help when faced with a particularly dangerous or difficult situation, these coalitions are temporary and are invariably established in the heat of the moment for a specific purpose. In contrast, the coalitions of monkeys and apes are often set up months ahead in anticipation of future need and rely on detailed social knowledge of both the likely opponents and the available allies within the social group.

Alliances are fundamentally important to the lives of all monkeys and apes in two major respects. First, the social group as a whole constitutes a collective alliance against the vagaries of the world at large. As with humans, the threats come from two separate directions: conventional predators like leopards, snakes and birds of prey on the one hand, and other members of your own species competing for access to life-sustaining resources on the other. Survival depends on having access to sufficient food and places of safety (especially safe sleeping sites). These are often in short supply, and there is invariably an undercurrent of competition among the local groups for control over these resources. Secondly, within the group, individuals encounter similar problems on the small scale. Merely by virtue of being forced to live together for protection, animals incur costs both through trespassing on each other's social space (so generating the stresses of social life so familiar to city commuters) and through competing for the best food items and the safest sites within a refuge. As the group forages across the habitat, the animals compete to avoid being on the exposed periphery where they are most at risk from predators. Having reliable allies helps to buffer you against all these minor stresses and prevents you being pushed to the edge of the group. The bullies are less likely to bully you if they know someone else will come to your aid; thieves are less likely to steal food from you when you are with your friends.

Monkeys and apes show a fine-tuned sense of the risks involved in supporting another individual against an aggressor, as well as an appreciation of the consequences of failing to support an ally in its time of need. While studying rhesus macaques on the Caribbean island of Cayo Santiago, Saroj Datta noticed that juvenile females were much less likely to support allies who were not their relatives when these were under attack from a member of a higher-ranking family than if the opponent was from a lower-ranking family. It seems that the monkeys were aware that supporting an ally against a higher-ranking opponent would simply result in both of them being trounced, especially if the opponent's family members came to its support. Better to hang back and avoid escalating the situation.

Since offending against an ally (either by failing to support it or by monopolizing a prize instead of sharing it) weakens an alliance and makes it less likely that the ally will reciprocate on a later occasion, monkeys and apes have a mechanism that is designed to repair the damage in such cases. Known as 'reconciliation', it usually involves some form of contact between the two parties in the aftermath of an altercation between them. In chimpanzees it may involve kissing, in baboons a touch or perhaps a brief episode of grooming. This serves to say 'sorry' and to patch up the relationship, to restore the *status quo*. So far as we know, these behaviours are unique to the monkeys and apes.

Primates need big brains, so the Machiavellian Intelligence hypothesis argues, because their social world is much more complex than the physical world. Whereas rules of thumb in the physical world remain more or less constant throughout an animal's lifetime (predators are always dangerous, rain is usually followed by a burst of fresh growth in the vegetation, flowers presage fruits a month or two later), the constancies in the social world occur only at the very highest level. A coalition with the biggest thug in the group is always the most effective, but the precise identity of the biggest thug can change from month to month, sometimes as a consequence of age-related changes in different individuals' strength and sometimes as a result of wholly adventitious events (including such factors as whether last week's chief thug happens to have been taken ill or left the group). Then there is the question of how to integrate the consequences that one's actions have for kin into the equation: one of the implications of Hamilton's Rule is that their interests have to be taken into account and added to or substracted from the equation as you go along. Not only is the equation itself very complex, but the constantly changing pattern of relationships in the group creates its own informational overload. There is no point in trying to get Jim to form an alliance with you against John if the two of them teamed up in an alliance of their own last week or if Jim is John's brother. Keeping track of who is in and who is out adds another dimension of complexity that does not exist in the physical world.

Primates, then, seem to have evolved specialized skills in the social domain. Indeed, there is some experimental evidence to suggest that, like Cosmides' human subjects, monkeys are good at understanding cues that relate to other individuals' behaviour but relatively poor at understanding ecological cues. Dorothy Cheney and Robert Seyfarth carried out a large number of tests on wild vervet monkeys living in Kenya's Amboseli National Park. They found that the monkeys were surprisingly unresponsive to a number of cues of ecologically relevant events: they ignored the presence of an antelope hanging in a tree (an unmistakable sign of a recent leopard kill, and hence of a leopard's presence) and of artificial and genuine python tracks in sandy soil. On more than one occasion, a vervet walked into a thicket while stepping nonchalantly across fresh python tracks, only to come racing out again a moment later after coming face to face with the snake itself. (The vervets alarm call vigorously on sighting snakes, for these prey on them.) In contrast, cues of social significance are recognized at once for what they are: the territorial call of the male from a neighbouring group played back from a loudspeaker placed in a different territory attracts instant attention (as though to say 'what the devil is *he* doing over there?'). So do a baby's cries of distress broadcast from a loudspeaker hidden in a bush. Not only do the monkeys become alert at once to the cries of the baby, but they invariably look first at the mother before looking at the source of the calls: they recognize that a baby is in distress, but more importantly they know whose baby it is and expect the mother to respond.

Monkeys, as we saw in Chapter 4, are especially good at learning social categories: they can distinguish between individuals and their dominance ranks, and recognize relationships like 'is friends with'. Seyfarth and Cheney were able to demonstrate that their vervet monkeys could recognize higher-order relationships like 'X is related to Y in the same way that Z is related to me.' This they inferred from the fact that after X had threatened Z, Z's coalition partner was significantly more likely to threaten Y (X's partner) within the next two hours than during a comparable period when no such attack on Z had taken place. We do not, of course, know quite what interpretation the monkeys place on the concept 'is related to': they could be interpreting it as 'Spends a lot of time near', 'grooms a lot with' or even 'has an alliance with' (in the sense of the reciprocation of aid in fights). The exact *meaning* is not especially important in the present context. The point of significance is that they behave as if they recognize the equivalence of the two relationships. That is quite a sophisticated comparison to be able to make, one that human children do not learn to make until they are five or six years old.

The social life of monkeys and apes appears to be unique in one other important respect, and this is the extent to which they use tactical deception in their relationships with each other. Tactical deception involves giving false information in order to gain an immediate advantage. Here's a typical example observed by Dick Byrne and Andy Whiten while studying baboons in southern Africa. Paul, a juvenile baboon, sits watching a young adult male digging up a succulent root. Roots require a lot of strength to uncover and pull from the ground, and a juvenile cannot manage it. Just as the male gets the root clear of the ground, Paul suddenly starts screaming and geckering at him. He is behaving as though the male had threatened to attack him. Instantly, his mother appears from behind bushes some distance away, takes in the situation at a glance and rushes to her son's aid. The astonished male, caught completely off-guard, flees before the onslaught, leaving his root behind. Paul picks up the root and begins to eat it, for all the world as though nothing untoward had happened. It seems as though Paul deliberately set the male up by behaving in such a way as to make his mother believe that the male was attacking him.

Did Paul *really* intend to deceive his mother into helping him get the root from the male? Or was it all a coincidence? Was Paul merely expressing his frustration at not being able to dig roots for himself, and his mother misunderstood his screams? This latter explanation, at least, seems implausible, for Paul was not giving vent to the usual baboon wails of frustration: he was screaming in just the way baboons normally scream when they are being attacked by a much more powerful animal. It seems more likely that Paul had learned that if you scream in this way, mum will come and rescue you, and it is but a small step from there to inferring that, if mum can rescue you from genuinely dangerous situations, you can exploit this to gain access to things that you wouldn't otherwise be able to do.

A behaviourist might argue that Paul is simply performing an action cued by a specific situation in the past, something he has learned by accident during many repetitions of the same kind of circumstance. But this is not a very convincing argument either, partly because deception (if it can occur at all) only works if you don't use it too often – it's the old problem of the boy who cried wolf once too often. So, at the very least, Paul has made some inferences about how his mother might behave on the basis of, at most, a handful of occasions. That juvenile baboons can make such inferences on the basis of very limited knowledge was suggested by an earlier study by Dorothy Cheney on another population of the same species of baboons. She found that when juveniles groom unrelated adult females, they devote most of their attention to the more dominant animals, presumably because these make better allies than lower-ranking individuals. If this is why they behave in this way, then their choice of prospective ally is based on very limited experience of the various females' effectiveness as allies. And most of that comes from noting how good these females are in aiding other juveniles rather than from assessing the aid that they themselves have received.

Tactical deception, it seems, is an important, if rare, phenomenon in the social lives of monkeys and apes. Here are some other examples from the literature just to make the point. The Swiss zoologist Hans Kummer spent many years studying hamadryas baboons in the Afar Desert in eastern Ethiopia. These baboons live in large groups of seventy to eighty animals made up from a number of small family units, each consisting of a single adult male, two or three adult females and their offspring. The males guard their females very jealously, punishing them with ferocious neck-bites if they stray too far away or allow another male to come too close to them. One day Kummer observed a young female spend twenty minutes edging her way from where she had been sitting beside her male to a large boulder a few metres away, behind which lay a sub-adult male. Eventually she managed to position herself behind the boulder in such a way as to groom the sub-adult male while having her head in full view of her own male, who presumably could feel confident that she wasn't misbehaving because he could still keep his eye on her.

The Dutch primatologist Frans de Waal spent many years studying the large chimpanzee colony at Arnhem Zoo in Holland. He noticed that the dominant males in the group were very jealous of their rights to mate with the females when these came into oestrus. Any younger male who deigned so much as to invite one of the females to mate was liable to be attacked, as was the female herself if she had been caught *in flagrante delicto.* The females, however, often preferred to mate with the younger males, and would sometimes actively solicit mating with them. When they did so, they would invariably encourage the males to move to a remote part of the large compound where they would be hidden from the dominant male by bushes. Similar tendencies for females to withdraw from the group when consorting with low-ranking males have been noted in Jane Goodall's famous population of chimpanzees at the Gombe National Park in Tanzania. More interestingly, perhaps, de Waal noticed that when the Arnhem chimps mated under such circumstances both the male and the female suppressed the loud post-copulation calls normally given at the climax of mating. Hans Kummer, who noticed the same effect in gelada baboons, has referred to this tendency to suppress conspicuous vocalizations so as to avoid giving the game away as 'acoustic hiding'.

In another study of captive chimpanzees, the American psychologist Emil Menzel showed a low-ranking chimpanzee where he had hidden a fruit in the large one-acre compound into which the animals were released each day from their night-time indoor cage. Having seen where the fruit was hidden, the chimp would be returned to the night cage and then the whole group would be allowed to enter the compound together. On the first few trials, the knowledgeable chimp would rush over to the hidden fruit and dig it up. But after it had been taken from him by a higher-ranking male two or three times, he soon learned to suppress his excitement on being released into the compound. He would avoid the location of the fruit and instead wander nonchalantly around some other part of the compound until the other animals had settled down. Once the dominant animals were preoccupied grooming or sleeping, he would steal over to where the fruit was hidden, recover it and eat it as quickly as possible.

If you spend time watching the social groups of monkeys and apes, as many of my colleagues and I have done in both the wild and captivity, the one thing that strikes you most is the extraordinarily complex social balancing act – a real-life soap opera – that is being played out among the animals. The animals seem to be constantly assessing where they stand and how things are going, responding to a slight tip in one direction by throwing their weight in elsewhere. They monitor not only what's happening in the rest of the group, but the effectiveness of their own strategies. Here are two brief examples to illustrate the point.

In his delightful little book *Chimpanzee Politics*, Frans de Waal recounts the story of how the young male Nikki formed a coalition with the former dominant male Yoeren to topple the new dominant male Luit. Neither the aging Yoeren nor the young Nikki were individually capable of defeating Luit in a one-to-one fight. But by acting in concert, they could. Nikki (who could easily trounce his aging ally) then ranked number one, Yoeren ranked second and poor Luit brought up the rear of the trio of males. The benefit of being the top-ranking male (the thing that made it all worth while) was the fact that you could monopolize matings with the females when they came into oestrus. The frequency of matings correlated directly with the males' ranks: Nikki gained the great majority, Yoeren was able to sneak the occasional one and poor Luit was once again excluded from all sexual activity.

For Yoeren, who had a few months earlier been the top-ranking male, the loss of privilege was clearly somewhat irksome and he made repeated efforts to mate with the females with whom he had previously been so intimate. Nikki, however, was not one to stand idly by wasting opportunities to mate, so whenever Yoeren made efforts to mate with an oestrous female, he intervened. The wily Yoeren's response was simply to wait until the next time Nikki and Luit got into a fight, and then he sat on the sidelines and refused to intervene on Nikki's behalf. The result, of course, was instant defeat for Nikki. It seldom took more than one such lesson to remind Nikki of his dependence on Yoeren, and then, for a while at least, he would tolerate Yoeren's dalliances with the females. The Japanese primatologist Toshisada Nishida has observed similar behaviour among the wild chimpanzees of the Mahale Mountains in Tanzania, and has given it the evocative name 'alliance fickleness'.

Primates spend much of their time monitoring very carefully what is happening around them. A nice example of this was provided by Barry Keverne during his studies of the reproductive biology of macaque monkeys. In many of these species, mating only occurs at the point in the female's menstrual cycle when ovulation is most likely to occur. Keverne was interested in determining the extent to which males used olfactory cues given off by the female to determine the proper time for mating. Some years earlier, Keverne and his colleague Richard Michael had been able to demonstrate for the first time that pheromones (sexual smells) existed in primates. This had been done by blocking the male's nasal passages with a removable coating of wax and then showing that, when his nose was blocked and he wasn't receiving any olfactory cues of her receptiveness, the male was no longer interested in mating with the female. In trying to repeat this procedure some years later, Keverne was puzzled to find that the frequency of mating remained the same whether or not the male's nasal passages were blocked. Looking more closely at the behaviour of the animals (rather than simply recording the end product of mating), he noticed that the female was now soliciting the male much more actively than she did when his nose was unblocked. It seemed that, so far from being a passive agent in the business of mating, the female constantly monitors the effectiveness of the signals she sends to the male. Normally the olfactory cues are so effective that she relies entirely on them. But when she discovered, in the experimental situation, that the message wasn't getting across, she quickly switched into another mode of communication and began presenting her rear end to the male in order to solicit his interest more actively.

In summary, despite millions of hours spent studying thousands of species of mammals, birds, reptiles and fish in the wild and in the laboratory, no one has yet produced any clear-cut evidence for tactical deception in any species other than a monkey or an ape. Nor is there any evidence to suggest that any species other than monkeys and apes uses coalitions for anything other than immediate self-defence. These appear to be two unique primate attributes and they are responsible for creating the extraordinarily complex societies that we commonly associate with primates.

As might be expected, it seems that the complexity of the judgements and decisions that monkeys and apes make in the course of negotiating their way through the maze of their social world translates directly into the size of the internal computer they have. Quite by chance, I recently discovered that there is a simple relationship between the relative size of the neocortex and the size of the social group in primates. The neocortex is the thin outer layer of the brain. Although it is just a few millimetres deep, it actually accounts for about 80 per cent of total brain volume in humans. It is the need to wrap this huge slab of tissue around a relatively small inner brain that causes the human brain to have its convoluted, deeply fissured appearance.

The neocortex is interesting for two reasons. First, it is the part of the brain that has expanded out of all proportion to the rest during the course of primate evolution. The neocortex is a new bit that has been added on to the basic reptilian brain by the mammals, but typically in most mammals it accounts for only about one third of the total volume of the brain. In the primates it becomes progressively larger, accounting for about 60 per cent of brain volume in the Old World monkeys, 70 per cent in the chimpanzees and 80 per cent in modern humans. Secondly, and more importantly, the neocortex appears to be the part of the brain where active thinking takes place.

The relationship between neocortex size and group size suggests that primates had to evolve increasingly large neocortices in order to cope with the increasingly demanding task of keeping track of the larger number of social relationships that occur in larger groups. Essentially, if bigger groups are needed to solve the ecological problems of survival and successful reproduction, this is only possible for primates if they first increase the size of the computer (i.e. the brain) used to keep track of all the relationships involved. This is a peculiarity of the way primates organize their social groups. Unlike other animals who simply tend to bunch together in rather loosely organized and unstructured groups, primates live in very tightly bonded groups that are organized into sets and sub-sets of special relationships, often built around kinship bonds that extend back for several generations. These relationships are held together through social grooming, but they are based on deep personal knowledge of the individuals concerned. It seems that it is the need to manage this constantly changing database and to draw accurate inferences about how other individuals will behave under different circumstances that makes it necessary for primates to have such large neocortices.

In order to be able to engage in tactical deception, it is, of course, essential that you keep track of who is currently 'in' with whom and how this will affect their behaviour. Being able to deceive someone else requires that you understand them well enough to be able to predict correctly how they will respond when you feed them false information. Dick Byrne has shown that the relative frequency with which instances of tactical deception have been reported for different species of monkeys and apes also correlates with the relative size of their neocortices. Chimpanzees, which have the largest neocortices after ourselves, seem to engage in tactical deception most often, followed by baboons and macaque monkeys (with the next largest neocortices), and so on down through the guenon monkeys of Africa, the spider and titi monkeys of South America, to the lemurs of Madagascar. It seems that the ability to exploit Machiavellian loopholes is directly related to the computing power that the animal can bring to bear on the problem.

The Mind's 'I'

One last component in this story concerns the psychological processes involved. In order to engage in the kinds of sophisticated social strategems that monkeys, apes and humans seem to specialize in, we need to be able to stand back from our immediate selves and ask how another individual is likely to respond when we behave in a certain way. At a very crude level, of course, all animals have to be able to do that in order to interact with each other at all. A dog has to recognize that another dog that is snarling with its lips drawn back and its tail held tensely up is likely to attack, and so dashing in to lick it all over the muzzle would not be the most appropriate response. But tactical deception needs something more than this. The actor needs to be able to see the world from another individual's point of view. In one sense that means being able to take a sufficiently detached view of the world that you can, as it were, step inside another individual's mind.

Of course, we can never know precisely what it is like to be another person: that has been the crux of another hoary old philosopher's chestnut known as the 'Other Minds' problem (Do other people really exist outside my own mind? Can I ever know for sure what another individual feels?). Nonetheless we can go some way towards imagining what it is like to be someone else. We do it partly by analogy (the noises and actions you make when you accidentally hit your thumb with a hammer are rather like the ones I make under the same circumstances, so it's likely that your internal feelings are similar to mine) and partly on the basis of what you tell me (your descriptions of your internal feelings using words that have a more or less agreed meaning between us). But being able to do this requires a very special ability, the ability to think reflexively about beliefs and desires. In effect, I have to be able to think about another individual's state of mind, to recognize that he or she has a particular belief about the world. And, furthermore, I have to be able to conceive (have a belief myself) that another individual has a belief about my state of mind (what he or she thinks I believe to be the case).

The philosopher Dan Dennett refers to this as being able to take an 'intentional stance'. In philosophical jargon, being able to think intentionally means being able to attribute beliefs, wants, desires and other similar states of mind to another individual. Individuals can have beliefs about their own states of mind ('I believe that I want X'), and this constitutes what is termed 'first order intentionality'. They can also have beliefs about someone else's state of mind ('I believe that you want X'), which involves second order intentionality. 'I believe that you think that I want X' is third order intentionality, while fourth order intentionality involves being able to say 'I believe that you think that I believe that you want X'. After that, as Dennett points out, things get a bit complicated, though humans can just about cope with sixth order intentionality if they think very carefully about it. In practice, most of our everyday social activity involves only first order or second order intentionality.

That the ability to think reflexively is essential for a competent social life is clear from studies of humans who suffer from the condition known as autism. Alan Leslie and his colleagues have shown that autistic children lack the ability to put themselves in someone else's place; as a result, they cannot see the world from someone else's point of view. Without this ability, autistic people lack the capacity to interact effectively in a social context: they are simply unaware of other people's feelings and needs. Instead, they treat the world of other humans as though it was the world of physical objects. They can cope quite competently with the physical world because it obeys simple laws that they can easily infer. Indeed, autistic people are often very good at routine tasks that depend only on learning facts: some of them can remember entire railway timetables, calculate calendar dates decades ahead, draw accurately from memory or play a piece of music that they have heard just once – tasks that normal individuals find all but impossible to do. But they cannot develop meaningful relationships with other human beings. Their relationships are superficial and take little account of another individual's wants or needs. They are locked into a world of simple physical laws.

Autistic children appear to lack another important psychological attribute, and this is the ability to engage in imaginary or make-believe play. They will not play with dolls, pretending they are people (when they *know* that they are not *really* alive); they will not pick up a toy telephone and hold a conversation with an imaginary person at the other end of the line; they never pretend to be asleep in order to play a joke on someone else. In short they live in a world that is absolutely real as it stands: they cannot conceive of the situation being other than exactly how it is. And that in turn means that they cannot lie.

Leslie has argued that the ability to engage in pretend play and that of being able to lie are closely related, appearing at about the same time during normal child development. They are, he believes, a reflection of the child's growing ability to detach itself from its personal perspective on the world so as to see the world from someone else's point of view. As such, this is not something we are born with in the sense that even babies do it: it is something that appears rather suddenly at about four years of age. It is not yet clear whether it is a capacity that develops naturally in the normal child come what may, or whether it is something that the child learns by a cumulative process of experience and practice. For present purposes, it probably does not matter too much either way. The important point is that children under about the age of four years are unable to distinguish other people's beliefs from their own; they cannot recognize that someone else can hold a belief about the world they know to be false. But from about four-and-a-half years onwards they can. Psychologists now refer to this as acquiring a 'theory of mind' (or ToM for short): by this they mean that a child starts to form theories (or, if you prefer, hypotheses or beliefs) about other people's states of mind. The criterion test for this is what is known as the false-belief test: a child who has ToM can believe that someone else believes something to be true that the child itself knows is in fact untrue.

One classic test for ToM involves a story about two dolls named Sally and Ann. Sally and Ann come into the room together. Sally has some sweets which she puts under the cushion on the chair. Sally then goes out of the room. While she is out, Ann takes the sweets and puts them in her pocket. Sally then comes back into the room. The child is then asked: 'Where does Sally think the sweets are?' A child who lacks ToM will not be able to distinguish his or her state of knowledge about the situation (the sweets are in Ann's pocket) from Sally's state of knowledge and will point to Ann's pocket. A child who has acquired ToM knows that Sally holds a false belief: she believes that the sweets are under the cushion, but the child knows that this is in fact not the case. This child will point to the cushion. The transition between these two states occurs at about four or four-and-a-half years of age, but it is preceded by a period of up to two years during which the child begins to attribute beliefs and desires to other individuals. At first these are likely to be its own beliefs and desires; later it discovers that other individuals can have beliefs and desires that differ from its own; later still, it learns that other individuals can have beliefs and desires that are false. Once it has discovered this, it can lie with the skill and persuasion of an expert. Autistic children consistently fail this test: they never develop ToM and they never lie.

However, it seems that all human beings who develop normally develop a theory of mind, and this is independent of cultural background. Cultural relativists, especially among social anthropologists, have repeatedly claimed that the way people view the world is influenced by the society or culture that they grow up in. However, it now seems that this is at best only partly true. A number of recent studies carried out by John Flavell and Paul Harris and their respective colleagues have shown that children from China and Japan, as well as Baka pygmy children from the Cameroon, all acquire ToM at about the same age as European and American children. Whatever nuances particular cultures may impose on people's perceptions of the world during the process of enculturation, they begin life with pretty much the same machinery, and it is a machinery that is especially sensitive to social phenomena.

Is ToM a uniquely human characteristic? It is clear that the vast majority of other animals do not possess ToM. Some doubt, however, remains about the species that are biologically closest to us. The closest of all is, of course, the chimpanzee, with whom we share something like 98 per cent of our genetic material and most of our evolutionary history. Sanjida O'Connell, a member of my research group, has recently managed to produce the first convincing experimental evidence that chimpanzees can attribute false beliefs to other individuals. It seems that they too have the capacity for ToM. But when she gave the same test to monkeys, they failed it: monkeys, it seems, cannot attribute false beliefs.

So although monkeys impress us by the complexity of their social strategems, they achieve these by using rather less sophisticated psychological processes than we (and perhaps chimpanzees) do. Robert Seyfarth and Dorothy Cheney have aptly described monkeys as being good ethologists but poor psychologists. They can read another animal's behaviour extremely well and use that to manipulate and exploit them; but they cannot imagine what it's like to see the world from another individual's perspective. Cheney and Seyfarth describe a wonderfully evocative example of this from the wild vervet monkeys that they spent nearly two decades studying in Kenya's Amboseli National Park.

One day the dominant male of their study group noticed a strange male hanging around in a neighbouring grove of trees. The stranger's intentions were quite obvious: he was sizing up the group in order to join it. If he succeeded, it was more than likely that he would quickly become the dominant male, for only the most confident and powerful males try to move between groups in this way. The incumbent male would quickly be ousted from his position of privilege and, like poor Luit at Arnhem Zoo, would be excluded from mating with the females. With the vervet equivalent of a stroke of genius, the male hit on the ideal ploy to keep the stranger away from his group. As soon as the strange male descended from the grove of trees to try to cross the open ground that separated his grove from the one in which the group was feeding, he gave an alarm call that vervets use to signal the sighting of a leopard nearby. The stranger shot back into the safety of his trees. As the day wore on, this was repeated every time the stranger made a move in the group's direction. All was going swimmingly until the male made a crucial mistake: after successfully using the ploy several times, he gave the leopard alarm call while himself nonchalantly walking across open ground. It was the monkey equivalent of the four-year-old who knows that if he denies eating the chocolates with sufficient vehemence adults will believe him, but who fails to recognize that the chocolate smeared around his mouth gives the game away.

This apparently trivial difference between ourselves and the monkeys (and indeed all other species of animals) has very important implications in two crucial, but related, respects. Only humans, it seems, can detach themselves sufficiently from their own view of the world to produce, on the one hand, fictional literature and, on the other hand, science and religion.

Fictional literature requires us to be able to think ourselves into someone else's mind, to be able to see the world from their point of view so as to imagine how he (not me, but he) would behave under a hypothetical set of circumstances. Our skill in doing this determines whether we can write something that another individual can empathize with, can recognize as being something he too has felt or experienced, or something that he can imagine himself as feeling or experiencing.

Science, too, requires us to be able to detach ourselves from our immediate perspective on the world, as does religion, but they do so in a different kind of way. They require us to be able to stand back and ask 'But *why* does that happen? Why is the world like that?' Where other animals accept the world the way it is, just as very young children and autistic people do, mature humans can step back from themselves and ask *why* the world is the way it seems to be. Young children, autistics, monkeys, other mammals, birds and fishes can learn generalizations about the world, but fail to take that one crucial step of going beyond the generalization to seek the reasons why they take the form they do. And this, it seems, is why humans, and only humans, have developed in any formal sense literature, religion and science. It also implies that humans will remain the only species to produce these three great products of the mind for the foreseeable future.

But not everyone is a Shakespeare or an Einstein. The great products of human culture are not the collective efforts of the five billion people who are currently alive. They are the products of a handful of particular individuals in whom the unique skills to which all humans are heir have been honed to a special sharpness. And herein, perhaps, lies the rub. To appreciate great science and literature probably requires the learning of skills that we do not naturally possess. We can, of course, appreciate Shakespeare at a certain level without any special training (in the sense that we can appreciate good music without being skilled musicians), but we can never appreciate it in the way that a trained writer or literary theorist appreciates it. Where the musical novice hears the orchestra as a single integrated sound most of the time, the professional musician hears each instrument's individual sequence of notes being interwoven with the others. The musician appreciates the subtleties of composition, the clever way an effect is created using certain notes at certain places in the score, but the novice simply hears a pleasant tune.

Although the same is true of science, science sometimes goes so far beyond the realms of everyday experience that the untrained mind cannot even begin to understand what the fuss is all about. We can play a single instrument's part and very quickly show a novice how the composer built up a musical sound. But none of us can begin to imagine what a sixteen-dimensional universe is like, or what infinity really means. At this point even the professional scientists abandon the attempt and confine their attention to the mathematics. The problems that this creates for popular science writing are the subject of the next chapter.

8
Science Through the Looking-glass

In another moment Alice was through the glass, and had jumped lightly into the Looking-glass room . . . Then she began looking about, and noticed that what could be seen from the old room was quite common and uninteresting, but that all the rest was as different as possible.

Lewis Carroll: *Through the Looking-glass* (1872)

Scientific information is essential, not only for the scientist. The politician, the entrepreneur and the public at large need to know about it too. The people in business find that neither the mass media nor the specialized scientific press are providing the information needed. General information is no longer enough, specialist information is only digestible for the learned. Who will bridge the gap?

Janny Groen, Eefke Smit and Juurd Eijsvoogel:
The Discipline of Curiosity (1990)

My theme so far has been that although the processes or methods of science are very much the stuff of everyday life, the theories that this methodology has produced are often very different from what common-sense experience might lead us to believe is the case. This ultimately gives rise to a serious problem about how we popularize science. Several issues hang on this, not least the fact that it is the public that ultimately pays for science in modern societies. If the layman does not understand what he or she is paying for and, worse still, lives in fear of the consequences that might befall from funding such research, then the future of science remains fragile: it will not take much to persuade the public at large that their happiness is better served by lowering taxes or the provision of more public amenities. At root, this is a problem about communication.

A Problem in the Public Domain

When Charles Darwin published his *Origin of Species* in 1859, he produced one of the classics of English literature. Written in elegant prose, its arguments lucidly laid out, the book was not only read by his zoological colleagues, but was also widely appreciated both by academics from other disciplines and by the public at large. Indeed, Fleeming Jenkin, who wrote one of the first reviews of the book, was a Professor of Engineering at Edinburgh University, not a professional biologist. Darwin's basic argument was easy to follow and much of the book was devoted to detailing examples that backed up his case. Many of these examples would have been familiar to his readers, either through personal experience (as with the many examples from animal breeding that Darwin used to support his claim for selection mechanisms) or through the flood of travel and exploration books that were then beginning to inundate the Victorian literary scene.

But this was far from true of physics by Darwin's day. Yet it once had been. Aristotle's works, for example, could be read and appreciated by anyone with a reasonable level of education. When Copernicus published his *De Revolutionibus* in 1543, the layman could follow the argument without too much difficulty. Even as late as 1632 when Galileo published his *Dialogue Concerning the Two Chief World Systems*, physics was still within the intellectual grasp of the educated layman, despite the smattering of mathematics it contained. But when Newton's *Principia Mathematica* rolled off the presses a mere fifty-five years later, things suddenly took a very different turn. At a stroke, science passed through the looking-glass and became incomprehensible to all but the professionals.

What had happened was that mathematics had become an essential component of the sciences. The arguments were raised on to a new plane whose understanding was only possible if the reader was familiar with the new mathematical techniques that included calculus and the beginnings of probability theory. Physics became opaque to the non-specialist.

In contrast, the biological sciences continued to remain accessible to the lay public for at least a further two centuries. Technical books like Darwin's were widely read. But no physics textbook published after 1700 ever featured on a publisher's best-seller list. The same process has, however, now begun to overtake biology. The first of the biological disciplines to pass through the looking-glass was genetics, which became fully mathematicized during the 1930s. Today a textbook in evolutionary genetics is too mathematical even for many biologists to understand. The chemically oriented disciplines within biology (like physiology and cell biology) began to suffer the same fate soon afterwards as their use of basic chemistry became more important.

Within biology, the areas that continued to remain accessible to the lay public were the behavioural disciplines like ecology and animal behaviour. But even these became mathematicized during the 1970s. As late as the mid-1970s, it was still possible to write a popular article on the behaviour of animals that any member of the public could empathize with. It was the behaviour familiar from your own back garden – birds squabbling over territories in the spring, singing in the dawn chorus and feeding their chicks. Matters changed imperceptibly but dramatically during the 1970s as biologists worked out how they could use the mathematical tools available in disciplines like population genetics and economics to study behaviour. Several important shifts in mind-set had been required to make this possible, as well as a spectacular growth in our knowledge of the natural history of a wide variety of species. As a result, it is now far more difficult to explain to the non-specialist just why behavioural biologists are interested in the problems that excite them. In order to say what's so interesting about the birds singing in the back garden, it is now necessary to preface your 2000-word article with a 10,000-word introduction to evolutionary theory at a level of knowledge that is far beyond that available to Darwin himself. The layman is left confused and floundering, the professional biologist tongue-tied.

The essence of this bind has been noted by the literary theorist Greg Myers. In his book *Writing Biology* Myers examines a number of popular science articles written by well-known biologists and compares them with the formal scientific papers they wrote on the same topic. He noted that the styles of these two kinds of papers are invariably very different. An article in an academic journal may deal principally with the nature of sex hormones, the research having incidentally been carried out on a particular species of snake. In contrast, the popular science article focuses on the snakes and their behaviour, with the sex hormones being incidental to the story, sometimes even taking third place behind the personality and experiences of the author.

Popular science magazines are commercial enterprises and their primary concern is to persuade their readers to buy copies. This places very demanding constraints on what stories they use and how they package them. The staff on these magazines therefore exert considerable editorial control over the content of the articles they commission from specialists. This results in a great deal of rewriting, the deletion of whole sections and the insertion of new material, not to mention a great deal of frustration for the scientists concerned. Often, to his dismay, the author will find his article (already rewritten to the limits of his tolerance to make it more reader-friendly) headed by an eye-catching title in the worst possible scientific taste. Some years ago, I wrote an article for the popular science magazine *New Scientist* about recent research on the way stress and excessive exercise can produce temporary infertility. When it appeared in print, my banal and rather boring title 'Stress and Infertility' had metamorphosed into 'Stress is a Good Contraceptive'. It was, of course, a decided improvement, but I remember feeling obliged to apologize for it when I sent a complementary copy of the article to one of the scientists whose work I had quoted at some length.

One of the examples that Myers discusses is an article written for *New Scientist* by the British evolutionary biologist Geoff Parker. Parker's research is concerned with the evolutionary processes of sexual selection and involved many hours spent lying on the ground peering closely at cowpats in order to observe the frenetic mating behaviour of dungflies. In writing an article for *New Scientist* on sexual selection, he concluded with a short section on the implications of his work for sexual selection in humans. This section was expunged from the final version by the editors. Myers rightly points out that it was cut not because it was regarded as too controversial (biologically speaking, it was in no sense controversial) or because it was in any way sensational, but because its inclusion meant that the reader 'would have to focus on the theoretical concepts underpinning the article rather than on the animals themselves' (Myers, 1990, p. 146).

In some ways Myers's comment does not go far enough: the problem is not just that it would have distracted the reader from a nice simple natural history story, but that it would have required at least double the length of the original article to explain how and why sexual selection might be expected to work across so wide a range of species as dungflies and humans in order to allay nervous readers' apprehensions about human beings apparently being treated as dungflies. Within the length constraints of the typical *New Scientist* article (2000 words), the content has to be fairly tightly focused on just one key point. It is always much easier to locate that point within the natural history of the species concerned than within the theoretical constructs of the science, if only because the one will be much more familiar to the average reader than the other. The problem for both writer and editor is just how much knowledge they can afford to assume in the reader.

What is interesting about nature to biologists is often rather different from what fires the average layman's interest. Parker is primarily interested in mating strategies and their impact on the evolution of behaviour via the process of sexual selection, and he has devoted a great deal of his professional life to developing mathematical models of the ways in which different mating strategies might evolve. The layman is more likely to be interested in the voyeuristic aspects of animal behaviour or in what the research can tell us about our own behaviour or that of the researcher – what was it about the researcher's psyche that made him or her engage in this peculiar form of voyeurism? The scientists' work will often be open to misinterpretation by the lay public because of this. For the layman, the only point of reference in Parker's article is likely to be the curiosity value of the fact that someone should spend so much time lying in fields staring at cowpats, not the fact that this particular research has made a more significant contribution to our understanding of the processes of evolution than almost any other since World War II. As Myers observes, the research is 'open to misinterpretation by nonbiologists who see [the scientist] as just a voyeur watching the sex lives of lizards and snakes . . . What comes across most strongly to a popular audience is simply the strangeness of the mating process in these species' (ibid., p.147).

A Metaphorical Problem

Part of the writer's problem is that an article written at too technical a level discourages readers who cannot understand what is being said. In the past, a major problem in this respect has been the use of specialized words and phrases. Scientists' use of specialized terms has, in part, been a consequence of the need to be sure that writer and reader were talking about exactly the same thing (essential if you are to check someone else's calculations or repeat their experiment). Scientists have often invented new words for these purposes, often with Latin or Greek roots. Such terms are often extraordinarily cumbersome. I did a quick trawl through the letter A in a dictionary of science and a conventional English dictionary: 79 per cent of everyday words consisted of three or fewer syllables, whereas 88 per cent of technical scientific terms had three or more (with as many as two in every hundred having seven or more syllables!).

However, this is only part of the problem. Individual disciplines also tend to develop their own styles of writing, often in abbreviated shorthand form. Most of this is intended for mutual convenience and to enhance speed of communication, but it can prove very confusing even for other scientists, as the following story illustrates.

In 1971 the editor of the prestigious *New England Journal of Medicine*, F. J. Ingelfinger, became exasperated by the near-incomprehensibility of papers submitted by immunologists. In order to highlight the problem, he had one of the papers rewritten by a professional science journalist in simpler English, and published the two versions side by side. Non-specialists responding to this experiment tended to approve of the rewritten version: it demonstrated that even an area as abstruse as immunology could be written in such a way that anyone could understand it. But the immunologists were not impressed: they complained that the rewritten version was harder to understand. In effect, the rewritten version might just as well have been a translation into French or Russian. Here and there you recognize a few familiar words but the exact meaning of a sentence is not clear because you miss the subtle nuances conveyed by the grammatical structure of the language. Immunologists had evolved a style of their own that made perfect sense to them. In this respect scientific disciplines do not differ from the everyday world, where the evolution of subculture dialects and styles of speech is a continuous process that gradually changes the overall content of a language and ultimately leads to fission into many mutually unintelligible languages.

The need to have words that refer to specific phenomena is extremely important in science: failure to do so results in unnecessary talking at cross-purposes, wastes people's time and holds up the advancement of science. Some disciplines, however, have made a conscious effort to avoid invented words, not least because these are difficult to learn and it is hard for novices to remember what they refer to. In disciplines like particle physics and evolutionary biology, where our everyday experience leaves us hopelessly unprepared, everyday words are often used in a punning way. If you understand the joke, then you'll never forget what the word is referring to.

A classic instance concerns the fundamental particles known as *quarks* (itself a word of uncertain etymology, but thought by some to be a direct borrowing of an obscure colloquial Austrian word meaning 'nonsense'). Quarks come in six varieties (collectively known as *flavours*) that include *top*, *bottom*, *up*, *down*, *charmed* and *strange*, each of which occurs in one of three *colours* (*red*, *blue* and *green*). Colour in this context has about as much to do with pigment as it does when musicians use the term of music (itself a metaphorical borrowing in its own right, of course). Similarly *up* and *down* are not only unrelated to each other as properties, but are also quite different from *top* and *bottom*. The terms *charmed* and *strange*, I suspect, reflect physicists' bemusement at the bizarreness of certain aspects of quark behaviour (and quantum physics in general). Remember that many of these terms were adopted to refer to theoretical entities before the discovery of the corresponding physical phenomenon – the *top* quark's existence was only confirmed experimentally in 1994. In fact, quarks might be impossible to see even in principle (all we can see is their consequences at the phenomenal level). In this case, the mathematics has forced physicists towards the idea that these entities exist.

Similar humorous usages of everyday terms have emerged out of evolutionary biology within the last two decades. Richard Dawkins's use of the term *selfish gene* (and its later echo in Francis Crick's *selfish DNA**) is a classic example of the genre. Other even more flamboyant examples include the Kamikaze Sperm Hypothesis (a reference to the wartime Japanese suicide pilots), the Beau Geste Effect (after the hero of a 1950s adventure novel), the Red Queen Hypothesis (from the character in *Alice in Wonderland*), a Hobson's Choice Strategy (after the proverb, itself based on an eighteenth-century Cambridge stable owner) and the Three Musketeers Effect (in honour of the book by Alexandre Dumas), all of which refer to bona fide biological phenomena. There has even been a trickle of papers with titles like 'Mate Choice in Plants' which are likely to strike the uninitiated layman as quite bizarre.

In all these cases it is, of course, possible to read the terms at two different levels. They can be taken literally (usually with the most peculiar results) or they can be understood metaphorically. Mathematicians, for example, refer to equations as being 'well behaved'. But they do not

* A large proportion of an organism's DNA (the genetic material we inherit from our parents) serves no obvious purpose and does not appear to be involved in the creation of new cells during the course of the individual's development. It seems to be 'junk DNA', to give it its alternative name. Crick (who shared the Nobel Prize for the discovery of DNA) suggested that it was viral DNA that had invaded the bodies of our remote ancestors at various stages during evolution and was hitchhiking its way through time for free using the bodies of the host organisms. It acts selfishly in the sense that it bears none of the costs of creating bodies in the face of natural selection, but benefits by being passed on from one generation to the next when the active DNA creates new bodies.

mean to conjure up images of equations sitting politely around the tea table waiting to be offered cakes and sandwiches; instead, they are saying something about the consistency and predictability of their properties: a well-behaved equation is one that does not produce nasty unexpected effects in some parts of its range. To understand the metaphor, you need to know the cultural background to the usage, as is true of all metaphors.

Similarly the term 'selfish' in Dawkins's concept *selfish gene* is not meant to be taken literally. Quite obviously genes cannot behave selfishly because selfishness is a moral property and it is likely that only humans can possess it. Dawkins is as well aware of this as anybody else. His point is an analogical one: genes behave (in the mathematician's sense, of course) *as if* they were selfish, as if they only looked after their own interests. In fact, of course, they do nothing at all: they are simply inert bits of DNA whose only functional capacity is to produce copies of themselves. But the processes of Darwinian natural selection act on genes in such a way that the effectiveness with which they are able to replicate themselves in the next generation *looks* like active choice. Dawkins is simply pointing out that when we ask about the evolution of behaviour (or anything else), we have to adopt a gene's eye view. His point is a purely cautionary one, a reminder that in evolutionary studies the accounting has to be done in terms of the number of copies of a given gene that appear in the next generation.

A literal interpretation of any of these terms would create nonsense (and, indeed, would often contradict the very point being made). Yet some commentators on evolutionary biology have expressed grave misgivings over the use of this kind of language. Howard Kaye, for example, castigates E. O. Wilson for his anthropomorphic use of language in his book *Sociobiology: The New Synthesis.* 'Wilson's sociobiology,' Kaye observes, 'is burdened by an equally misleading cybernetic model of organisms. Such gratuitous mechanizing and systemizing appear to endow the commanding genes with consciousness and will and the evolutionary process with rationality and purpose' (Kaye, p. 109). Doubtless it does to a non-biologist, but observations of this kind are inclined to strike evolutionary biologists as naive, at best. How could these terms possibly have any everyday connotations, for they are defined rather precisely, often in mathematical terms?

We ought at this point to ask what it is about particle physics and evolutionary biology that leads to this kind of metaphorical use of ordinary language. It is conspicuous, for example, that these usages are far less common in chemistry and anatomy. Why should this be so? Does it mean that physicists and evolutionary biologists are less scientific than chemists and anatomists? Or is it that chemists and anatomists are just boring people?

The answer is neither of these. What unites physics and evolutionary biology (and, incidentally, computer science) is that they deal with phenomena that everyday experience does not equip us to talk about. These phenomena are so far outside our normal range of experience that we simply do not have words readily available to describe them. In contrast, chemists and anatomists deal with phenomena for which the conventional mechanistic terminology of everyday language is quite suitable. The simple mechanistic analogies of everyday life break down in particle physics, while evolutionary biology looks at the world in a way that is more or less at right angles to the way we normally do. (We are used to characterizing people's and animals' behaviour in terms of motivations and intentions, whereas evolutionary biology focuses on function measured in terms of genetic consequences.) In addition, both of these disciplines are highly mathematical and many of the phenomena being referred to are mathematically defined entities that we cannot actually see. Lacking a suitable language from everyday experience, physicists and biologists have been forced to use everyday language (it is after all the only one we have), but they use it in a metaphorical sense, often signalled by a deliberate punning element. Understandably this makes life hard for outsiders who see words they recognize but fail to realize they are being used in unfamiliar ways. However, the gain in both the clarity and comprehensibility of papers for the professionals more than justifies the costs incurred from the occasional mistake by the uninitiated.

Another reason why particle physics and evolutionary biology may be unusual is that, in both cases, powerful new paradigms have recently opened up particularly exciting and productive avenues of research. The world has turned out to be even more complex than we had thought, yet the principles that underlie this complexity have proved to be elegantly simple. In each case the new theoretical paradigms have proved to be extraordinarily powerful, allowing us to predict and explain the 'behaviour' of the world in quite remarkable detail. The excitement and euphoria that this has produced has no doubt enhanced scientists' sense of mastery over their art. But more importantly, I think, it has generated an immense sense of wonder, itself reflected in a heightened respect for the remarkable complexity of the natural world and the elegance of the principles that underlie it. Physicists, in particular, often refer to the sheer 'beauty' of the mathematical theories of modern physics.

This is not to say that scientists do not sometimes use words in their everyday sense, or that they never fall into the kinds of traps that critics have worried about. Most descriptive scientific terms undoubtedly begin life as reflections of common folk wisdom or personal experience. In most cases additional connotations are probably imported from everyday usage during these early stages. But if the term begins to be used as a serious technical term, then it invariably becomes increasingly more refined in its definition. In the process, it loses most of its everyday connotations.

Richard Dawkins provides us with a nice example of this process in his book *The Extended Phenotype*. He points out that the biological term *fitness* has progressed through five separate stages of meaning since it was first used in the 1860s by the philosopher Herbert Spencer. The first was in the rather vague everyday sense of physical fitness (as in the catch-phrase 'survival of the fittest'). Then in the 1930s it was defined more precisely as a technical term in genetics as the rate at which a given allele (such as the character brown eyes) is replicated relative to other alleles at the same locus (e.g. the character blue eyes). Fitness thus ceased to be a property of the individual and became a property of the gene. Later, during the post-war period, the need to assess the genetic impact of an animal's behavioural decisions resulted in fitness being redefined operationally to mean reproductive success (the number of offspring an individual produces in a lifetime). Later still, in the 1960s, the puzzle of trying to explain the evolution of altruistic behaviour (such as alarm calling or non-reproductive worker-bees) led to the concept of 'inclusive fitness': a gene's fitness included not only the reproductive output of a given animal, but also a discounted proportion of the reproductive outputs of all that individual's relatives because relatives share some of their genes by virtue of descent from a common ancestor. However, it turns out that inclusive fitness so defined is only an approximation to 'true' fitness, which is now sometimes termed 'neighbour-modulated fitness': this calculates a pooled value for the success of all the copies of a given gene in the population at any one moment.

This sequence of changes illustrates how a term's meaning can change over time as it comes to be defined ever more precisely in the light of new discoveries. Old usages are not so much wrong as imprecise. They are adequate as approximations within the range of contexts for which they were developed. They remain adequate for those purposes even now. But other problems may require a more refined definition. Clearly there is considerable scope for confusion here among those who are not aware of the changes in meaning that have occurred.

On the other side of the fence, scientists frequently complain that people in the social sciences and the humanities deliberately seem to be trying to confuse matters when they write articles and books. They often use unnecessarily complex terminology and an impenetrable writing style that seems designed to confuse and obfuscate rather than clarify. One of my biological colleagues once remarked that the difference between lectures in the sciences and those in the humanities was very simple: in the sciences the important thing was what you had to say, not how you said it, whereas in the humanities what you said (if anything at all) was irrelevant, but how you said it appeared to be everything.

The physicist Dick Feynman related an anecdote that illustrates this point rather nicely. He was once invited to join a prestigious conference of the great and the good from many different disciplines who were to spend a week closeted together to sort out one of the world's problems. Such intensive workshops became rather fashionable in the 1960s and 1970s (a phenomenon that even found its way into literature in the form of Arthur Koestler's novel *The Call Girls*). A sociologist had prepared a written paper ahead of time and circulated it among the discussants so that they would have a clearer grasp of what he wanted to say. As Feynman puts it in his own inimitable way:

> I started to read the damn thing and my eyes were coming out: I couldn't make head nor tail of it! . . . I had this uneasy feeling of 'I'm not adequate,' until finally I said to myself, 'I'm gonna stop, and read *one sentence* slowly, so I can figure out what the hell it means.' So I stopped – at random – and read the next sentence very carefully. I can't remember it precisely, but it was very close to this: 'The individual member of the social community often receives his information via visual, symbolic channels.' I went back and forth over it, and translated. You know what it means? 'People read.'
>
> (Feynman, p. 281)

Joseph Schwartz provides us with another example in his book *The Creative Moment*. The French social theorist and philosopher Pierre Bourdieu once wrote:

It is clear that critical discourses and displays can break the doxio relationship to the social world which is the effect of correspondence between objective structures and previously internalized structures only insofar as they objectively encounter a critical state able by its own logic to disconnect the pre-perceptual anticipations and expectations which form the basis of ahistorical continuity of the perceptions and actions of common sense.

With the help of novelist and literary theorist David Lodge, Schwartz translates this into simple English as: 'Genuinely original critical thinking is possible only in a revolutionary climate such as *les événements* in France in May 1968' (Schwartz, pp. 171–2).

And, in case you don't believe me, here is another example. In her book *Primate Visions*, the American historian of science Donna Haraway attempted to interpret the history of primatology (the study of primate behaviour and biology) in terms of recent political changes. In considering the recently published book *Primate Societies* (intended to be a state of the art compendium of our understanding of primate behavioural biology), Haraway found it to be:

an examplar of a widespread groping in the 1980s western biopolitical and other cultural discourse for ways to narrate difference that are as deeply enmeshed in feminism, anti-colonialism, and searches for non-antagonistic and non-organicist forms of individual and collective life, as by the hyper-real worlds of late capitalism, neo-imperialism, and the technocratic actualization of masculinist nuclear fantasies.

(Haraway, p.373)

'Gee,' observed primatologist Meredith Small in her review of Haraway's book, ' . . .[and] we thought it was just a text book.'

Schwartz points out that many people find abstract language of this kind appealing precisely because it has a distinctly poetic quality. It has an intrinsic beauty of its own. Dressing simple ideas up in obscure language makes them seem more profound than they really are. Moreover, like all good poetry, it can be read as saying almost anything you like: it can mean all things to all people.

We sometimes forget just how important the reader is in the literary process, for no matter what the writer may say, it is the reader's ability to make sense of what he or she reads that ultimately determines whether communication is possible. This was demonstrated rather dramatically by the publication in 1985 of a collection of poems and short stories entitled *The Policeman's Beard is Half Constructed* that had been composed by a computer program whose acronym was RACTER. In composing the stories and poems, the program chose successive words at random from its dictionary. If the chosen word fitted grammatically, it kept it and moved on to the next word-position in the sentence; if it didn't, it tried another word. The sentences produced by the program are, of course, meaningless nonsense, but the human reader can extract quite meaningful interpretations from them! The success of the experiment can be gauged by the fact that the book even received quite approbatory reviews in leading newspapers.

This kind of game appeals to the Postmodernist fringe in the humanities because it achieves Postmodernism's avowed aim of breaking the traditional one-way relationship between writer and reader. We have always assumed that the writer's aim is to get the reader to grasp the particular idea that the writer has in mind. But Postmodernists insist that the reader has the right to interpret a given author's work anyhow he or she likes. If you and I arrive at completely different interpretations of Shakespeare's *Macbeth*, so be it: that is our privilege. In the ideal Postmodern world, the writer should write in such a way as to make possible radically different interpretations of his or her work by different readers. Unfortunately, when it comes to understanding how the real world works, we have to do better than this: science is necessarily a communal activity and we have to be able to communicate our findings to each other in order to be able to build the collective jigsaw.

Science and Society

This brings us back to the problem that is fundamental to the whole issue of public (mis)understanding of science: public perception of what constitutes science is invariably rather different from what scientists actually do. To the average layman, science is about new discoveries. There is still much of the Victorian explorer–naturalist in the layman's perception of science. Part of the reason for this, of course, is that new facts are easy to report. They mean something to the average newspaper reader or television viewer in a way that a development in some esoteric theory almost certainly does not. And this has very much coloured the way science is handled in the media.

By and large, regular science reporting is a relatively new activity for newspapers. Hitherto, newspaper coverage has largely been confined to major events – the discovery of a new planet, the detonation of the first nuclear bomb, the discovery of a gene for some disease or a new fossil hominid unearthed from an ancient African lake bed. Regular science columns (pages would be too generous a description) began to appear in newspapers only within the last decade or so, largely, I suspect, in response to the success of television science programmes like the BBC's 'Horizon' series and the American 'Discovery' channel. Yet even these count for a tiny proportion of the column inches published each week in the national press. Of the 660 hours of TV offered on a typical day (23 April 1996) on the 33 channels available in the Boston area, just 10h 50m was devoted to science programming (1.6% of total viewing time). A significant proportion of this was devoted to wildlife or health programmes, and most of it was scheduled after midnight and on cable. The newspapers fared no better. Of the 1,585 column-inches of text (excluding obituaries, advertisements, letters, pictures and cartoons) printed in the *Boston Herald* on the same day, 50 (3.2%) were concerned with science. In addition, the paper carried a 12-page supplement on health which raised its science coverage to 4.5%, though again most of that dealt with health issues. On the other hand, just 2 of the 2,486 column-inches published in the upmarket *Boston Globe* on the same day were devoted to science. Even allowing for this paper's regular Monday science slot (roughly 200 column-inches in the week's total of 15,000), science coverage accounts for only around 1.3% of its text.

When science does feature in these papers, it tends to be in the gee-whizz vein. It rarely involves any background analysis of the kind that even the most trivial political event would invite, and it seldom tries to place the discoveries being reported into their context within the relevant discipline. This is, I think, largely a consequence of two failings in the media.

One is the fact that, until recently, very few journalists have had any kind of background in science. Most journalists learned their trade hands-on, working their way up through local papers on to the nationals, while those that came via the graduate-entry route invariably held degrees in the humanities. Aside from the few specialist magazines like *New Scientist* and *Nature*, science journalism was a cinderella corner for those not good enough to make the political and economic pages or edit the arts page or the book reviews. Editorial control by non-scientists has inevitably meant that science reporting had to compete on the same footing with conventional news stories. Media wisdom has it that news must have impact and, especially, human interest to sell papers. But when science tries to compete with the social antics of the great and the not so good, it has only limited chances of success.

This, of course, is not entirely the fault of the media hacks, who are in part simply responding to people's buying habits. But it is worth noting that the media's assumptions about what their readers want is not always very reliable. The great shock of the past decade for book publishers has been the existence of a covert market in popular science. Serious science books for the mass market have long been viewed as a non-starter, but the tail-end of the 1980s witnessed some astonishing successes once publishers built up the courage to take the plunge. Stephen Hawking's book *A Brief History of Time*, which dominated the publishers' best sellers list for months on end, is but the most spectacular of these.

The other common failing in the way science is handled in the media stems from the fact that most of the media, both written and aural, suffers from an absurd level of time pressure. I am only too well aware from my own experience working as a professional freelance journalist that too often pieces have to be written too fast to permit adequate research. This invariably leads to shortcuts of convenience: it's more important to fill the page with something readable, almost irrespective of its content, than to do proper justice to an issue when that means investing time into background research. Even then journalistic research tends to involve asking an accessible expert rather than reading any of the primary literature. Inevitably the journalist's own uninformed opinions become more intrusive. I gave one such example in Chapter 1 in a reference to the discovery of the gene for maleness. Here are some other examples that make the same point.

I remember some years ago circulating the science correspondents of the major daily papers with the programme of a meeting I was organizing for one of Britain's smaller science societies. The topic of the meeting was the use of primates in biomedical research and its effect on the conservation of wild populations. It was an issue of some significance from both the ethical and the commercial points of view. I was anxious to generate some public discussion of the issues on both sides of the argument. But only one journalist turned up: he came for one talk only, and that was inevitably the one that looked like being the most sensational. The piece he wrote in the following day's edition of his paper reflected that choice.

On another occasion a journalist from one of the national mass market daily papers rang me in search of a quote that he could use to support the particular view he wanted to promote. I spent half an hour explaining to him why this view was scientifically untenable. He listened patiently, but the only thing he wanted to know was whether I could suggest the name of a scientist who would support his view!

More recently the science correspondent of one of the country's leading daily newspapers declared that he was unable to find anything of interest to say about a three-day meeting of the British Psychological Society which he had covered; psychology, he surmised, was in a state of decline. So far as I am aware, he acquired all his information by talking with a self-selecting handful of speakers in a quiet corner, and not always in a particularly informed way. Had he taken the trouble to attend some of the lectures, he would have learned about some genuinely extraordinary research.

Such sloppiness does not inspire great faith. But it is worrying: no matter how much journalists may plead that they are merely purveyors of information and not opinion-formers, the printed word does carry its own aura of respectability and truth. If a journalist on a highly respected national newspaper asserts that there is nothing of interest or importance to report from the shop window of a discipline, then that must inevitably influence the views of politicians and those who hold the purse-strings of science.

To be fair, of course, journalists should not be blamed for all the ills that befall the public presentation of science. There are a large number of science journalists who do an excellent job despite the constraints on their freedom of movement. More importantly, perhaps, scientists must themselves bear some of the responsibility for the poor press they have received, for they have often been reluctant to dabble in popular science writing. Indeed, until recently, writing popular articles was considered to be a very questionable activity, and it certainly would not be counted as part of a scientist's productive output. Even when they have dipped their toes in the popular market, scientists have not always done so with great success. So much so, in fact, that they have acquired an reputation for lamentable style and cultural boorishness. Bryan Appleyard, for example, has argued in his book *Understanding the Present* that the discipline of science inevitably destroys the individual's aesthetic sense. Science, he wants to suggest, acts as a kind of universal thought-police that produces grey shadows of humanity who lack that wonderful sense of individualism that makes us human. The worry is that, if Appleyard is right, not only does this produce acultural (or even anti-cultural) boors but it makes for very boring journalism. Science will grind itself into a pulp of its own making. By failing to excite the minds of the rising generation, it will either turn itself into a sterile religion or die from its own apathy.

But how true is it that a training in science so stultifies the mind that scientists are incapable of appreciating art and high culture? What of Brian Appleyard's claim that science inevitably destroys our aesthestic sensibility?

The Philistine in the Laboratory

It is easy to show that scientists are not necessarily cultural boors, either by constitution or by training. A surprisingly large number of scientists have made important and lasting contributions to cultural activities as different as music, art and literature.

Many people will be aware that a significant proportion of professional science fiction writers like Isaac Asimov, Fred Hoyle and Arthur C. Clarke are professional scientists. Asimov is a chemist and Hoyle a physicist, while Clarke was largely responsible for the original concept of communications satellites. Others will know that the novelist C. P. Snow was a chemist who lectured at Cambridge University before World War II. Later he became a scientific advisor to the British government, helping to formulate its post-war investment in public science. Rather fewer people will know that the Russian Nobel Prize-winning novelist Alexander Solzhenitsyn was also a professional scientist. After obtaining a degree in mathematics at the University of Rostov, he taught physics and chemistry for some years before turning to writing. Primo Levi, the Italian novelist, was also a chemist.

Of course, not all scientific novelists achieve such illustrious heights. But many others have produced substantive literary works. After a decade or so writing successful non-fiction books on the human face of science, the American astrophysicist Alan Lightman, for example, has recently published his first novel, *Einstein's Dreams*, to considerable critical acclaim. The late John Treherne managed to combine lecturing in zoology at the University of Cambridge with a burgeoning literary career in his later years: he wrote two successful books of historical biography (one on the American gangsters Bonny and Clyde), as well as two novels. Marie Stopes, the turn-of-the-century pioneer of birth control clinics, was both a scientist and a published poet and novelist. Then there is the young German evolutionary biologist Volker Sommer: so far he has produced three novels, a book of poetry, two books on the cultural symbolism of the rose and another on festivals, as well as literally dozens of scientific articles and books.

Nor have scientists been absent from the field of music. The great Russian composer Alexander Borodin (widely considered to have been technically the most innovative composer of the second half of the nineteenth century) was head of the Chemistry Department at the St Petersburg Medical Academy. Not only was he a pioneering teacher (he introduced the first medical courses for women in 1872), but he also made significant contributions to science. His best-known discovery was the aldol condensation reaction (aldols are polymerized derivatives of alcohol), and he devised an analytical test for urea that was used well into the twentieth century. Borodin was a member of a small but influential group of Russian chemists based in St Petersburg. His colleague and friend Dimitri Mendeleyev (who occupied the chair of chemistry at St Petersburg University) gave us the Periodic Table of the chemical elements that adorns every school laboratory throughout the world. Other musical scientists include the nineteenth-century Russian military engineer César Cui (composer of many songs, orchestral pieces and operas) and the modern Estonian minimalist composer Arvo Pärt (a radio engineer by profession).

But the prize in this category must surely go to Ludwig Ritter von Köchel, whose name will be thoroughly familiar to anyone with an interest in classical music despite the fact that they will know very little about him. Born in 1800, Köchel took a doctorate in Law at Salzburg University; then, while tutor to the imperial Austrian household, he became an internationally respected authority on both the botany and the geology of Europe. He built up a substantial geological collection during his travels (subsequently bequeathed on his death in 1877 to the city museum in Vienna), and left us with several plants bearing his name. It is this same Köchel who provided us with the official catalogue of Mozart's works, a task that took the better part of a decade to complete and which would almost certainly have been impossible without his earlier experience in cataloguing plants and minerals.

On the more artistic side, the physicist Dick Feynman (another Nobel Prize-winner) was an accomplished artist who sold and exhibited his drawings in California during the 1960s. Others have made art a more serious part of their lives. The biologist Jonathon Kingdon, for example, makes his living out of selling his paintings and sketches of animals, while at the same time producing acclaimed technical books, including an enormous five-volume monograph on the anatomy and biology of the mammals of Africa.

E. B. ('Henry') Ford, who was Professor of Genetics at Oxford in the 1960s (and singlehandedly founded the sub-field of ecological genetics), was also a keen amateur archaeologist of some distinction. He excavated Cornish *fogous* (prehistoric earth-houses) and published an erudite monograph on the ecclesiastical antiquities of Oxfordshire. Then there is Noam Chomsky, the founding father of modern linguistics and a major intellectual force in the cognitive sciences generally: he can also lay claim to credentials as an historian, having written what is widely considered to be one of the most erudite analyses of the Spanish Civil War.

This is a far from exhaustive trawl through the spare-time cultural activities of scientists. It completely ignores the many scientists who paint, write poetry and perform music as amateurs, not to mention those whose interests in art, architecture, literature and music take them to art galleries and theatres at least as often as the average layman. Not long ago I took part in a small conference of biologists in Madrid. To my casual knowledge, more than half of those who attended took time out to visit Madrid's famous Prado art gallery or the Queen Sophia Museum of Modern Art (where Picasso's *Guernica* painting hangs) or attended concerts during the four-day meeting. I wonder how many participants at a comparable meeting in the humanities would have taken the trouble to visit Madrid's science museum or its zoo?

Thus the claim that scientists as a group are cultural philistines simply isn't true. The average scientist is probably as literate and artistic as the average graduate in the humanities. Instead, we ought to be asking this question: is the average professional in the humanities as scientifically literate as the average scientist is culturally literate?

It would, of course, be unfair to castigate everyone in the humanities for being scientifically illiterate. There are striking exceptions, and these serve as a reminder that an interest in the arts is not necessarily incompatible with an interest in science. Here, then, are four scientifically literate members of the humanities community.

The Edwardian artist and children's writer Beatrix Potter was a keen botanist and naturalist in her early years and even wrote a paper on the germination of the spores of the Agarinacae (a family of fungi) which was read before a meeting of no less a body than the Linnean Society.* The

* Sadly, she was not allowed to read the paper herself because women were not allowed to attend meetings of the Society at that time.

composer Edward Elgar dabbled in chemistry, while Robert Simpson, whose musical credits fill half a column in *Who's Who*, is a keen amateur astronomer who lists astronomy as his main recreation and has the distinction of being a member of the British Astronomical Association. David Bolter had a distinguished early career in the classics, but then switched disciplines mid-stream to become a computer scientist. His book *Turing's Man* was a deliberate attempt to set out the common ground between science and the arts in order to allay the deep-rooted suspicions that non-scientists often have of science.†

As an aside on the relationship between science and music, it's worth noting that the work *The Light* by the American minimalist composer Philip Glass was commissioned by the University of Minnesota to commemorate the centenary of the Michelson–Morley experiments. In 1887 two members of the University's Physics Department, Arthur Michelson and Edward Morley, carried out a series of ingenious experiments that proved once and for all that space is not filled with an ether, as received theory then supposed; in doing so they paved the way for Einstein's Theory of Relativity which was published just two decades later. Science, it seems, is not so remote as to be unable to recognize the artistic in its own achievements.

Yet, if anything, these are the exceptions that prove the rule. Far too many of those who consider themselves to be artistically well-informed are scientifically illiterate. But if never having listened to a Bach cantata or seen an Australian Aboriginal rock painting is an astonishing cultural lacuna in one's life, then so is not being interested in how stars are formed or how genes produce bodies in that greatest of all miracles on which life itself so precariously depends. Science is, by any standards, one of the most astounding achievements to date of the human mind: how is it possible for some people to be so woefully ignorant of so great a part of our cultural heritage?

† Alan Turing was the English mathematician whose work in computational mathematics in the 1930s laid the essential theoretical groundwork for modern computers: modern digital computers are a direct realization of the purely conceptual Turing Machines that he invented.

9
The Open Society Revisited

The main philosophical malady of our time is an intellectual and moral relativism.

Karl Popper: *The Open Society and its Enemies* (1962)

To the relativists, one can only say – you provide an excellent account of the manner in which we choose our menu or our wallpaper. As an account of the realities of our world and a guide to conduct, your position is laughable.

Ernest Gellner: *Postmodernism, Reason and Religion* (1991)

Up to this point, I have argued that science works as well as it does because it makes us ask questions of nature. We are forced, under the constant pressure of asking 'But *why* does this happen?', to confront and challenge nature in such a way as to make it reveal its secrets. In this chapter, I shall be asking the converse question: what happens when science does not have such a free hand, but is constrained in what it can do by social or political interests?

In his book *The Open Society and its Enemies*, the philosopher Karl Popper argued that attempts to constrain science within a political straitjacket are invariably disastrous. His main concern was to counter what he perceived to be the spreading influence of a totalitarian form of Marxism among intellectuals during the 1950s. But his point remains as relevant today when other forms of fundamentalism threaten to constrain science. Attempts to impose intellectual frameworks that are wholly derived from the local culture are bound to interfere with the natural processes of science. If the theoretical structure is preconceived and wholly impervious to change, then progress cannot occur because the theories remain inviolate to criticism. Paradoxically, then, in suggesting that science is socially constructed, the relativists have in fact identified the very thing that makes productive science impossible.

Through a Glass Darkly

During the 1950s and 1960s the sociology of knowledge was dominated by the so-called 'Mertonian Paradigm' (named after its instigator, Robert Merton). Merton's argument was founded on a defence of unfettered science. He argued forcefully that scientific institutions act so as to maximize their own productivity in terms of the purposes they set themselves. Left to themselves, he insisted, scientists produce good science because science itself is a self-regulating rational process. The so-called 'Strong Programme' developed as a reaction against this 'Whig' view of science during the 1970s, and is associated in particular with the Edinburgh School and with, among many others, the names of Michael Mulkay, Mary Hesse and Steven Shapin. The basic argument adduced by the 'Strong Programmers' among the sociologists of knowledge was that, so far from producing progress and new knowledge, science, as much as any other kind of human intellectual activity, simply reflects the local culture's particular interests and concerns at any given time. In their view, science is the prisoner of its times, reflecting the partisan interests of the culture within which it is embedded. 'Science,' wrote Wolf Lepenies, a leading German social theorist, 'must no longer give the impression it represents a faithful reflection of reality. What it is, rather, is a cultural system, and it exhibits to us an alienated interest-determined image of reality specific to a definite time and place' (Lepenies, p.64).

Analogous views had been expressed by the so-called Frankfurt School in Germany during the 1930s and 1940s. Its leading lights, Max Horkheimer and Theodor Adorno, had argued that science had become 'mythical', creating a form of blind obedience to, and worship of, superior intellect that was reminiscent of pre-Enlightenment religion. In doing so, they argued, science had become a social tool of powerful élites that were anxious to enforce their stranglehold on society through a totalitarian domination of state institutions. Their ideas reflected the Marxist assumption that science is heavily influenced by social and political preconceptions. Although very much a prisoner of their own troubled times, the Frankfurt School's views naturally resonated with the Durkheimian tradition in the social sciences. Among social anthropologists, for example, the prevailing view during the 1960s began to be that traditional religious systems were not attempts to explain natural and supernatural phenomena but rather externalizations of social and political forces internal to society (see Chapter 3).

Steven Shapin and Simon Schaffer, in particular, have argued a very strong case for the claim that the norms of behaviour in science are designed to promote science, so that to accept them uncritically, as they argue the Mertonians had done, is to accept science's own view of itself. In effect, they said, we would be accepting science's right to set its own agenda, irrespective of whether or not that agenda was desirable when seen from the wider perspective of society as a whole.

The typical approach that the cultural constructionists adopt is to take a particular scientist (such as the biologist Charles Darwin or the chemist Robert Boyle) and try to show that his conception of the problem(s) he became interested in was largely a product of the society in which he happened to live. Darwin's interest in the problem of evolution was thus largely determined by contemporary society's concerns with explaining the origins of life in the aftermath of the Enlightenment's devastating attack on religion. Darwin, like many of his contemporaries, was caught up in a battle royal between the rearguard action of organized religion and the new liberal humanism that traced its origins back to the French Revolution. Similarly, it might be noted, the mechanisms of evolution that Darwin chose to emphasize were based on the liberal economic theories that dominated mid-nineteenth-century thinking. The *Origin of Species* is, after all, full of references to Malthus's tract on over-population and the breeding-for-profit experiments of the Agricultural Revolution. Thus, we are led to infer, Darwin's ideas were a product of the social and economic milieu in which he lived. We are but a step away from claiming (as many have indeed tried to do) that the Theory of Evolution is so riddled with mid-Victorian class-based values that it forms part and parcel of the capitalist programme. Wittingly or unwittingly, so the argument runs, Darwinism's modern advocates are engaged in implementing capitalism's agenda.

Most scientists, faced with this kind of analysis, are inclined to exasperation. Where else, they will ask, are you to get your hypotheses from? They can only come from the experiences of the individual scientist or from the existing theoretical structures of the discipline, and either way they reflect the scientist's cultural milieu (in the one case social, in the other case scientific). But this does *not* mean that they stop there, for that would be to imply that scientists do not actually follow their own methodological prescription of testing the hypotheses that they have proposed.

In some ways, what is odd about the sociologists' claims is precisely that they take absolutely no account of history despite making a great play on the historical basis of the analysis. To be sure, the analysis *is* historical in the sense that it seeks to place the scientist or the particular problem under investigation into its historical and cultural context. But there history ends. Any sense of the temporal *process* of history, that history is, quite literally, a dynamic story that runs and runs, seems to be totally missing. The sociologists' conception of history is a purely static and contextual one: it is not historical at all.

So does modern evolutionary theory reflect the parochial interests of the contemporary world? The proverbial man-in-the-street is, of course, now reasonably familiar with the notion of genes and the process of evolution. Yet the layman's grasp of the details of evolutionary theory is invariably lamentable. Notions that were eradicated from biology half a century or more ago ('evolution for the good of the species', 'nature red in tooth and claw', 'survival of the fittest [i.e. the strongest]' to name but three) pepper the discussion because they are still widely believed by the lay public (and even by the non-biological disciplines within the sciences) to be substantive components of evolutionary theory – despite the fact that most of these have nothing at all to do with Darwinian biology. More often than not, the realtionship between science and popular culture is the reverse: modern culture has acquired many of its ideas secondhand from science, often in emasculated form.

In many of these cases, the sociologists' problems seem to be due to their complete lack of familiarity with the technical content of the scientific disciplines they have been trying to study. Unable to grasp the scientific arguments, they have invariably resorted to superficial analyses of the social relationships between the members of the relevant scientific community. In their book *Higher Superstitions*, Paul Gross and Norman Levitt offer hilarious exposés of the scientific naiveties perpetrated by some widely respected sociological studies of science. Some of them are so bizarre as to begger scientific belief.

An alternative approach sometimes adopted by sociologists of knowledge has been to examine the public writings of scientists on the grounds that they reflect scientists' underlying attitudes to their work. Steven Woolgar takes this approach in his analysis of the Nobel Prize acceptance speech of an astrophysicist. His concern was to interpret the text of this speech as a political statement in which the history of an event (such as a discovery) is reinterpreted in the light of subsequent history and what is now politically expedient. This is signalled, in some sense, by the importance of the occasion (the award of a Nobel Prize). This, he argues, encourages the creation of a very conventional myth designed to 'explain' how things came to be as they are now and to bolster the status of the key players in the story.

Woolgar's analysis seems to me yet another example of the kind of problems that sociologists can get themselves into when they fail to appreciate the nature of the context in which particular texts are written. Whatever the importance of myths for the promotion of science as a publicly funded activity, they play no role at all in the *production* of science. Certainly no one in science takes them very seriously. At best they are part of the PR of science that keeps the money flowing in for what is often very expensive research. To compress many years' work and much wrangling and argument (not to mention detours down what later turned out to be blind alleys) into a matter of a few thousand words in such a way as to make both a coherent story *and* a story that is entertaining for the layman is no mean feat. It simply cannot be achieved without doing serious injustice to the actual history of events. What emerges is always a caricature, a distillation of a few key strands of the story, often heavily edited to isolate the points that, with the benefit of hindsight, were most central to the developing saga.

In a classic example of this, the historian Donna Haraway shows, in her book *Primate Visions*, how the National Geographic Society wove a fabric of myths around the public persona of Jane Goodall and her work on the chimpanzees of Tanzania's Gombe National Park. These myths were unquestionably very important in promoting public interest in studies of wild primates, and were probably single-handedly responsible for persuading governments and independent funding agencies to underwrite the costs of research in this area for several decades. They also probably played a seminal role in attracting many young people into this particular field of study, providing the launch-pad for a thousand PhDs. But those myths contributed hardly at all to the development of science in this area. Our present understanding of primate societies and their evolution owes little to studies of chimpanzees, and even less to the work of Jane Goodall herself, despite the genuinely important and very substantial contributions that she and many others like her have made to our store of knowledge about the behaviour of monkeys and apes. And it owes nothing at all to the feature stories that appeared in the pages of the *National Geographic* magazine.

A failure to understand how scientists work seems to have been responsible for other problems of interpretation by sociologists and historians of science. In science, a great deal of time is spent arguing about the validity of particular empirical tests and in trying to define a 'fair test' (see Chapter 2). This has frequently surprised many of those who expected science to run on rigorously rational Popperian lines – negative results should lead to the immediate rejection of the hypothesis under test, not to an argument about the validity of the test. Sociologists have inevitably recognized this (quite correctly) as being a process of negotiation. Unfortunately they then seem to have made an inferential leap and presumed that these kinds of negotiation are the same as more familiar processes like pay-bargaining and diplomacy, where the object is to find some mutually convenient compromise. Presumably you decide what you can reasonably get away with from an experiment, and then double its value in order to be able to negotiate your colleagues down to something close to your initial position.

If this really were true, then science would indeed have no external validity: it would simply consist of arbitrary compromises between sets of conflicting beliefs. On this interpretation, the conclusion that light behaves both like a wave and a particle is merely a diplomatic solution to an otherwise insoluble problem that threatened to tear physics apart in the last century. Alas, this is not how the process of negotiation actually works in science. The aim is not to find some mutually acceptable compromise (though this no doubt happens from time to time), but to identify a test that will uncompromisingly confirm one hypothesis at the expense of all its rivals – or, as in the case of light, shows us precisely why a given phenomenon should present two different faces to us. If scientists cannot agree on what counts as a 'fair test', then they cannot do science.

It would, of course, be churlish to condemn all sociologists of knowledge: even the sociology of knowledge is not without its heroes. Prominent among these is Michael Lynch who has himself castigated sociologists of science for totally failing to understand the nature of the phenomenon they have been concerned to study. Lynch argues that if you want to understand what scientists do, what the significance of their conversations are, how their writings should be interpreted and so on, you must become as familiar with their subject-matter as the scientists themselves are. The sociologist has to become a competent laboratory scientist. Lynch is, of course, simply emphasizing what every good anthropologist knows to be true: participant observation (immersing oneself in an alien culture in order to understand it from the inside) is the only possible way to proceed. And there is no question but that science *is* a foreign culture, as much for the sociologist as for the layman. For Lynch, the sociologist's programme has been at best hypocritical: it has done little more, he argues, than foist its own simple-minded preconceptions on to the fabric of science. Bravo for common sense!

Robin Horton has been another lone voice crying in the wilderness – in this case, the wilderness of contemporary social anthropology. He has vigorously defended the view that traditional religious systems function as a primitive form of science in pre-modern societies. 'One of the principal intellectual functions of traditional African religions,' he writes, 'is that of placing everyday events in a wider causal context than commonsense provides' (Horton, p.56). Religion in traditional society serves to provide a coherent framework that makes sense of those phenomena that cannot easily be explained by simple cause–effect processes. Horton contrasts the way religion functions in traditional societies with the way it does in modern post-industrial European societies: in our modern societies, science has tended to replace the explanatory framework function of religion, leaving it only the moral and psycho-emotional spheres to deal with. One consequence of this is that it is possible for people to be both scientists and religious. In this respect, he argues, it is religion in our societies that is odd; in traditional societies, religion permeates everything a person does, the way he thinks about the world as well as his behaviour towards others. Of course, religions can and do function in this way even in our own societies: most major religions provide some kind of metaphysical framework which gives coherence to the universe and guarantees our place in it, but generally speaking this function is much reduced.

Building on their Durkheimian roots, the social constructionists among the anthropologists have tried to argue that religious systems are merely attempts by the secular powers to manipulate the members of their society so as to enforce their preferred social order. Horton is scathing in his criticism of leading social anthropologists like the late Edmund Leach who argued precisely this line in his landmark monograph *Political Systems of Highland Burma*. Horton argues that accounts such as Leach's convey an astonishing sense of unreality: it is, he insists, obvious to anyone who has lived in a traditional society that religion works to enforce social and political conformity precisely because everyone *believes* it to be a true explanation for natural events. It is perhaps relevant that Horton has spent his academic career teaching at universities in West Africa, where he has been safely distanced from the more corrosive intellectual influences that have developed in contemporary social anthropology.

Before bringing this section to an end, let me be quite clear about what I am saying. I am not claiming that scientists' ideas are never influenced by the common culture of the day; nor am I saying that they never impose their political or other preconceptions on the world they study. Of course they do, and probably much more often than we would wish. Mistakes of logic are made, the evidence fudged and the results of tests fiddled, all in a desperate effort to preserve a theory that an individual has committed most of his or her life to. Who amongst us wants to end a lifetime that has been devoted to the pursuit of a theory with nothing but failure to show for it? Scientists are human, and fall prey to all the foibles to which humans are prone.

But this is not the end of the story. Two things help to prevent malpractice on a massive scale becoming universal among scientists. One is the inherent scepticism of colleagues who would prefer to see their own alternative theories triumph. For better or for worse, it is precisely the competitive individualism of science that has saved it from the cult-forming fate of most religious and political sects. At the same time, however, rampant individualism remains less conspicuous in science than it does in most aspects of everyday life because scientists form a community with a common purpose. That sense of communality leads to much sharing of information and ideas, even with complete strangers. The second factor that preserves the integrity of science is the uncompromising intransigence of nature: you cannot force nature to behave in accordance with your pet theory, no matter how hard you try. Providing you relentlessly pursue the business of generating new predictions to check and double-check your theory and test these rigorously at each step in the chain, nature will force you to admit defeat by driving you inexorably into a corner where the evidence and your interpretation no longer match up. It is in the failure to keep asking questions that science degenerates into religion.

The Purse-Strings of Science

A more serious claim, perhaps, is the suggestion that science is in the pocket of its paymasters, that its continued funding depends on its willingness to reinforce the economic and political *status quo*. It is easy to point to well-known examples that seem to back up this claim. The development of accurate clocks was spurred on by the British Royal Navy who offered a cash prize to anyone who could produce a chronometer that would enable their ships to locate their positions more accurately at any place in the world. Their concern was not with disinterested science, but with the need to move armies around the world in order to protect Britain's newly acquired empire and its trading base. Archimedes invented many of his ingenious engines of war (not to mention his famous test for the purity of gold) at the behest of his mentor, King Hieron of Syracuse. Chinese astronomy achieved remarkable heights in the first millennium BC because the imperial court wanted to be able to foretell the future well enough to control its unwieldy empire, and good descriptions of star patterns were essential for good astrology.

It goes without saying that science depends on the generosity of the public purse. This is true whether the purse-strings are held by national governments (as they are today) or by individual benefactors (as they were in Newton's day). Either way, of course, it is ultimately all of us that pay, for that purse is simply the surplus wealth that you and I created by our labour and someone else has seen fit to redistribute. Science, like organized religion and the arts, can only exist where there is sufficient surplus wealth to allow some individuals to parasitize the rest: you cannot engage in intellectual activities of this kind if you also have to scrape a living. To produce art and science of the highest quality, you need to be able to devote every waking minute of your day to it, just as professional tennis stars and musicians have to spend a large proportion of their time practising the skills they need in performance. The rest of society has to be willing to strike such a deal with you; and for that, it expects some return.

This much is surely obvious from the fact that neither large-scale science nor institutionalized religions (i.e. those that support professional priesthoods) are found in hunter–gatherer societies. Only agriculturally based economies produce sufficient surpluses to purchase the specialized arts and crafts on which these activities depend. Even so, genuine explanatory science seems to have taken off only once, in Europe and the Near East during the first half-dozen centuries either side of the birth of Christ. Toby Huff has argued that it was the unique combination of the philosophical, legal and religious presuppositions of the Middle Eastern world that allowed science (as the challenging of conventional folk wisdom) to flourish in the West in a way that it was not able to do elsewhere. The Greeks' curiosity about the world was surely one important element in this, but an equally important factor was that their slave-based economy produced a leisured class who, unlike the Roman patricians that followed them, had no empire or trading interests to absorb their energies. It is surely no accident that modern science has achieved the heights it has in precisely those economies that have generated sufficient surplus wealth to create a leisured class.

However, the fact that the State underwrites science need not always mean that bad science (or bad art for that matter) will result. We need only point to the many beneficial advances that have been made under the economic and political pressures of war. The development of immunology as a science and of plastic surgery, as well as radar, rocketry and jet engines, not to mention computers, all owe their origins directly to the exigencies of World War II. Without the desperate need to find solutions to some very immediate problems of survival, none of these technologies would have been developed as soon as they were.

The space race provides another example of the same phenomenon. Whatever their reasons for doing so, the US and Soviet governments invested unheard of sums in, to quote the comedian Tommy Lehrer, putting some clown on the moon. The pressure to produce workable solutions to the formidable technological problems of space travel was enormous. After all, men's lives were at stake, never mind national reputations. The result was unprecedented advances in technology. We eventually benefited from the spin-off from these programmes in quite unexpected ways, including the development of new materials – the teflon used to create non-stick saucepans being one notable everyday example. But by far the greatest benefit was to come from the need to miniaturize computers. Deep space missions could not be undertaken using earth-bound computers because of the time it took messages to travel between the spacecraft and ground control. The success of this enterprise led first to a revolution in the office, then in newspaper and TV production and finally in the arrival of the home computer.

Among the many by-products of all this is the fact that some of the smaller cargo ships can now put to sea with crews as small as nine where once they would have required several dozen men. Gone are the traditional six watches of deck and engine room crews, the teams of navigators, radio operators and helmsmen that ensured the ship's safety at sea; even the cook has gone, replaced by a freezer and a microwave oven. A single officer controls everything from the bridge, courtesy of a computer and a satellite navigation system, in permanent contact with his head office via a celestial phone link. The minimum crew has been reduced to a captain, two officers, two engineers, two helmsmen and two deckhands (in each case, one on duty, one asleep). The laborious business of calculating the ship's position using sextant and compass at midday, of swinging the lead to measure the depth of the seabed when entering port – all these age-old nautical rituals are now a thing of the past, replaced by Star Wars technology.

So far from driving the progress of science by demand, the reality is that the burgeoning empires have, more often than not, merely provided sufficient surplus capital to enable science to be funded without raising too many public objections. After all, it took two world wars to persuade most governments that investment in basic science was a worthwhile use of taxpayers' money. The commercial exploitation of the resulting discoveries has often subsequently been taken up by governments and industry (though often many decades later), but theirs has invariably been an exploitation after the fact rather than a demand-led intervention. Scientists have generally done what they have been paid to do and rushed on as fast as possible to use what time and money they have left to explore the really interesting (but often wholly 'useless') problems.

Archimedes provides us with a nice example of just this. Probably the most renowned intellect of his day, his fame rested on the many technological inventions he created for his political masters. His engines of war (among them a large mirror said to have been capable of setting fire to the ships attempting to breach the harbour defences) so terrified the Roman armies that the defenders of Syracuse were able to successfully prolong the siege for more than three years. Among his less martial devices was a mechanical apparatus that allowed his mentor, King Hieron, to move a fully laden ship single-handed and the device known as Archimedes' screw (a contraption for lifting water through a tube by means of a rotating screw). But he himself regarded all these ingenious machines as of such trivial interest that he left no written accounts of them, despite the fact that he scribbled voluminously, producing ten mathematical treatises on geometry and physics that have survived and at least another six lost works that we know about from references in other sources. The things he considered sufficiently interesting to write down concerned such esoteric questions as how to calculate mathematical functions or why bodies float. Hardly the kind of stuff guaranteed to bolster faltering Mediterranean city-states in the third century BC. In contrast, we know about his technological inventions only from third parties who witnessed them, and then often in such wildly exaggerated form that it is all but impossible to reconstruct his machines. Only Archimedes' screw survived the mythologizing process because versions of it continued to be used in the irrigation of Egyptian fields right up until modern times. But it was his esoteric mathematical work rather than his engineering devices that were to stand the test of time, for they laid the foundations on which Newtonian physics was later to be based.

The origin of our modern atomic theory of chemistry provides another example. We owe this to Antoine Lavoisier, a businessman, tax collector and sometime geologist who, in 1775, was appointed by the French government to oversee the production of gunpowder at the National Arsenal. The latter half of the eighteenth century was a period of more or less continuous war in Europe, with France heavily implicated on one side or the other (and sometimes both at once). The production of good-quality gunpowder was a matter of supreme national importance. The way in which gunpowder burns (and hence the explosive force it creates behind a projectile) is determined by the ratio of its three ingredients (saltpetre, charcoal and sulphur). Lavoisier used his accountant's penchant for balancing accounts to carry out very careful quality-control analyses – probably the only occasion when accountants have proved to be of any real value to science. As a result he completely reorganized the manufacturing process at the Arsenal so as to produce the best-quality gunpowder in Europe – not to mention the Americas, where French gunpowder found its way to the revolutionary armies of the nascent United States and materially contributed to the defeat of their British colonial masters in the War of Independence.

But during his time at the Arsenal Lavoisier did a lot more than simply try out different combinations of the three ingredients. He became interested in what happens when substances burn, and in particular the combustion of sulphur and phosphorus (the latter is not one of the constituents of gunpowder). According to the chemical wisdom of the day, a substance called phlogiston is given off when anything undergoes combustion. Lavoisier's meticulous quantitative experiments had revealed that rather than getting lighter when they burned (as ought to be the case if they were losing something), burned matter gains weight when all the by-products are added back together again, while the air around them is progressively reduced in volume. 'About eight days ago,' he wrote in 1772, 'I discovered that sulphur in burning, far from losing weight, on the contrary, gains it; it is the same with phosphorus.' Later he demonstrated that water consisted of a mixture of hydrogen and oxygen and that it was oxygen that was removed from air during combustion to give rise to an oxidized compound of the element being burned. Lavoisier had completely turned the theory of chemistry on its head.

But the French government had not asked Lavoisier to produce a theory of chemistry. Had his gunpowder not been of such extraordinary quality, there would surely have been words in high places. Not only did the government have no interest in a new theory of chemistry, it did not even need one to produce high-quality gunpowder: that was simply a matter of getting the mixture right, something that Lavoisier had already managed to do by simple cookbook trial and error. The theory of chemistry emerged as a by-product of Lavoisier's practical work, aided and abetted by a bit of curiosity. The French government's part in the whole business was limited to providing Lavoisier with a salary plus some space down at the Arsenal to set up a laboratory.

If nothing else, these two examples ought to remind us that technology gets superseded, but esoteric fundamental science that lacks any immediate benefit provides for the advances of the future. My point is a very simple one. No amount of throwing money at the wall will, of itself, solve problems at the frontiers of science. Answers can be developed to the more immediate problems of technology in this way, but science needs time and motivation. Scientists will take up the challenge only if they themselves have an intrinsic interest in the problems concerned. It's all a question of psychology. Scientific research is too frustrating and boring most of the time to sustain the interest of anyone who is not fired by an overriding fascination for a particular problem.

The Subversion of Science

The secular State's influence on science has not always been as benign as it was in Lavoisier's case. All too often State intervention in science has proved a disaster. As it happens, Lavoisier provides us with an archetypal example of the adverse effect that the State can have on scientific progress. It was Lavoisier's bad luck to have been a tax collector for the monarchy in the days immediately before the French Revolution. It did not endear him to the *citoyens* of the new regime and in 1794 he was tried before a revolutionary tribunal and sent to the guillotine along with his father-in-law the same afternoon. He was fifty-one years old. Who knows what more he might have achieved had he lived. In the words of the mathematician Joseph-Louis Lagrange, uttered the following day: 'It required but a moment to sever that head, and perhaps a century will not suffice to produce another like it.'

Lavoisier was, of course, neither the first nor the last scientist to fall foul of a temporal power. Archimedes himself was put to the sword in 212 BC by an over-eager soldier in the aftermath of the capture of Syracuse by the army of the Roman general Marcellus – the latter's orders to spare the sage notwithstanding.

However we view this, one thing is clear. In no sense can economic or political interests be said to have dictated the course of science over the last two or three centuries. This much is clear from the fact that when public interest does attempt to direct science, the science it produces is rarely especially successful. I will offer just two examples that illustrate different aspects of the problem.

The history of Soviet biology is, perhaps, the best-known example. The problem here was essentially that, in the aftermath of the Russian Revolution, the person who managed to claw his way to the top of the hierarchy in biology, T. D. Lysenko, held Lamarckian views on the processes of inheritance. At that time the long-running debate between the Darwinian and Lamarckian forms of evolutionary theory was drawing to its final conclusion in Darwin's favour, although there remained convinced Lamarckians in many parts of the biological establishment in both Europe and America.* It remains unclear whether Lysenko's rise to political prominence in the Soviet hierarchy owed more to contemporary Marxist preferences for a Lamarckian view of evolution (it made the

* Lamarck had argued for the view that evolutionary change came through practice – the inheritance of acquired characters. The blacksmith's children inherited their father's muscles because the daily hammering that built up his muscles altered the heritable component (we would say 'genes') that he passed on to his offspring. Half a century later Darwin argued for the inheritance of traits that were the product of accidental changes in the genetic material, with natural selection acting to preserve those variants that proved to be more successful whenever they were able to reproduce more often. Only those individuals with the genes for big muscles made successful blacksmiths, so only these variants from the population at large were passed on to the next generation, irrespective of how practice might affect the muscles of the blacksmith himself during his lifetime. A further half-century later still, the new science of genetics confirmed that Darwin had been right. What is now known as Weismann's principle, or the Central Dogma, states that the relationship between the genotype (the collection of genes an individual inherits from its parents) and the phenotype (what it actually ends up looking like) is a strictly one-way process: genes influence phenotype, but not the reverse.

Marxist programme of self-development possible in a way that the Darwinian view seemed not to do – though this in itself was an unfortunate misinterpretation of Darwin's ideas) or to Lysenko's own promises to cure Russia's chronic wheat shortages. Either way, from about 1929 onwards, he used the power of the Soviet State to banish Darwinism from Russian biology and to institute a programme of biological research based on a theory that had by then been discredited.

It was not until the 1950s, after the death of Lysenko's mentor Stalin, that Soviet biology was able to emerge from what the historian Peter Bowler has described as its 'nightmare'. Political xenophobia had reinforced Lysenko's personal authoritarianism by preventing Soviet scientists from having any intellectual contacts with the West. While Western agriculture had been able to exploit to considerable advantage the new discoveries of Mendelian genetics, Soviet agriculture lagged increasingly far behind. The result was an agricultural disaster. While Russian biologists tinkered with plant breeding, the Russian peasants starved.

The great irony, indeed tragedy, of this story is that at the time of Lysenko's rise to power, the Russian geneticists working under the leadership of Sergei Chetverikov were well ahead of most of their European and American rivals. During the 1920s Chetverikov had been able to demonstrate that a species harbours a vast reservoir of concealed genetic information in the form of recessive alleles.* It was the rediscovery in the 1930s and 1940s of much of Chetverikov's work by geneticists of the calibre of R. A. Fisher, J. B. S. Haldane and Sewall Wright that proved to be instrumental in overthrowing the then accepted view that evolution was a consequence of the frequency of genetic mutation, so paving the way for the synthesis of the Darwinian and Mendelian theories into the modern theory of Neo-Darwinism. Sadly, Chetverikov was among the earliest casualties of the Lysenko regime and, despite the fact

* An allele is a particular version of the gene for a character. The gene for eye colour, for example, has two common alleles, one for brown eyes and one for blue. One allele of a given gene is normally inherited from each parent, with the two together determining the offspring's eye colour. Under simple Mendelian genetics, one allele can be suppressed by the other (as the allele for blue eyes is by the brown eye allele in humans). Such alleles (sometimes also confusingly called genes) are referred to as *recessive*: they can only produce their effect (blue eyes) when present in the offspring in a double dose. Nonetheless, they covertly transmit the capacity to produce that effect down through the generations. Many human genetic diseases are of this type.

that he lived until 1959, nothing of any consequence emerged from what had been one of the most dynamic and scientifically innovative laboratories in the world.

A less well-known example is the effect that Islamic religious fundamentalism had on the rise of Arabic science during the thirteenth and fourteenth centuries. Islam's insistence that everything of any consequence had already been written down in the Koran made it impossible for the philosopher–scientists of the Arab world to debate openly any of the newly discovered dimensions to science. Broaching new frontiers was considered blasphemous because it implied that God, in dictating the Koran, had done an incomplete job, and any implication of imperfection was deemed to be an insult to the Almighty – something that was punishable by instant death. The result, as Toby Huff has pointed out in his book *The Rise of Early Modern Science*, was the active suppression of the developing sciences in the late medieval Islamic empires of the Near East and North Africa just at the point where they were leading the world in the embryonic sciences of chemistry, astronomy and physics (see Chapter 3). The classic case of this was the twelfth-century Moorish philosopher al-Rashid, better known in the West as Averroës. Al-Rashid managed to maintain a substantial output (including his authoritative commentaries on Aristotle) while under the protection of the politically powerful caliph Yusuf of Cordoba. But when al-Rashid fell out with Yusuf's successor Yusuf al-Mansur (mainly thanks to the pressure from fundamentalists anxious to control the insidious influence of the independent thinkers), his career came to an abrupt end. Al-Rashid's death in 1197 marked the beginning of the end for Arabic liberal science in the West.

Fundamentalisms of any kind, whether religious or political, are a serious hazard for science. Indeed, any attempt to redefine the nature or concerns of science places the growth of knowledge at risk. In recent years a number of attempts to constrain the free range of science have emerged out of the self-styled radical philosophies developed by some feminists and, more recently still, by the politically correct movement.

The feminist critique of science has argued repeatedly that science is sexist and that its aims and concerns need to be redrawn along radical feminist lines. It is not always clear what this means in practice, but in his book *Scientific Realism and Socialist Thought* Andrew Collier has suggested that there are at least three different possible interpretations to this programme.

One possibility is that sexism (i.e. the deliberate exclusion of women from science by men) is acting as an obstacle to scientific advance in much the same way as the Church's geocentrism was an obstacle to Galileo's new scientific programme. But, as Collier points out, this is not sexist science but sexism disrupting the natural processes of science. Widening the constituency of science to include more women will surely lead to more and better science (not least because it might prompt women scientists to explore different questions), but it will not produce a different *kind* of science.

A second possibility is that the two sexes genuinely differ in their emotional responses, such that men make better scientists than women because they are able to achieve a greater degree of objectivity towards their subject-matter. But to condemn science because it might reveal that sex differences are built into the human psyche is a bit like smashing the microscope that reveals the bacteria that cause illness. However nice it may be to do away with differences between the sexes, denying science will not make them go away. In any case, this is all somewhat presumptive, since we are as yet a long way from demonstrating that any such sex differences in scientific ability actually exist. What differences do exist in this respect are much more likely to be a consequence of differences in motivation that largely reflect women's greater interests in the business of parenting rather than intrinsic differences in scientific ability.

An alternative interpretation is that the whole project of science, with its empiricist base, is sexist and that some other form of knowledge (for example, phenomenology or intuition) would be superior. Unfortunately the realities of life are such that introspection cannot tell us how or why the world works; it can only tell us about our own private thoughts. A whole range of essential matters of great practical significance would automatically be placed beyond the realm of human understanding. If it really is the case that men are better at objective practical science and women better at the more introspective arts, then the only conclusion we can draw is that women will never do good science or men good art. But this claim is so implausible it hardly merits further consideration.

Like all attempts to constrain science within a particular political or religious framework, these 'radical' critiques of science are unlikely to help us deal with the practical problems of day-to-day survival. By impeding the free flow of ideas, they risk stultifying scientific progress. This is not to deny that women scientists might open up new and interesting questions because of their particular interests and experiences. But those questions must remain firmly embedded within the framework of science and subject to its conventional methods if they are to be useful for the advancement of knowledge. Were it otherwise, science would necessarily be a closed book to other cultures. Yet members of other cultures, from Jabir ibn Hayyam to Chandrasekhar, have made substantial contributions to modern science. Science is, above all, an open forum: it will consider any ideas that look promising, irrespective of where they come from. There is every reason to encourage women to see science and technology as an interesting and fulfilling career, but we will not produce better science or save the world by putting them in some kind of scientific ghetto.

The Hidden Persuaders

If State control is bad for science, the lack of public understanding has to be worse. We humans are just too susceptible to the herd instinct. Religions are so much a part of our lives that we commonly fail to recognize how easily we fall prey to the will of small influential groups. But the results are to be seen all around us in the fanaticism and militancy generated through the ages in the name of religion.

The dangers inherent in these kinds of movement lie not so much in their particular beliefs, but in their demands that adherents surrender their independence of mind. Buddhism, with its emphasis on self-help and personal achievement, remains the one shining exception to what is otherwise a near-universal rule. Once we humans are reduced to being mere believers, we seem to be incapable of making balanced judgements. The stranglehold that organized religions have had over the human mind through the centuries have given rise to such extraordinary incidents as the 'Witches of Salem' in the USA and the 'Devils of Loudun' in France during the seventeenth century, the medieval crusades and the *fatwas* of late twentieth-century Islam. In all these cases, people have believed so passionately in a particular proposition that they have been willing to put to death those with whom they disagree. It is the staggering facility with which we humans seem willing to surrender ourselves to others rather than think for ourselves that is probably the most frightening aspect of our own behaviour.

One particular example of this phenomenon is worth exploring in more detail. It is the Romantic movement that developed in Germany towards the end of the eighteenth century as a reaction to the rationalism that had come to dominate academic and public thinking during the

Enlightenment of the preceding century. Its driving force lay in an attempt to reinstate the emotions and the immediacy of sensory experience in the forefront of human existence. Its main thrust was self-consciously anti-science, emphasizing the poetical and the spiritual over the empirical and the rational. Nature was viewed as an organic entity rather than the mechanical system described by contemporary science. It emphasized the mysterious at the expense of the knowable and eulogized (indeed romanticized) the medieval period which Enlightenment philosophers had viewed with a horror otherwise reserved only for the primitive and the ignorant.

The leading intellectual light of the fledgling movement was undoubtedly the German Idealist philosopher Friedrich von Schelling. Schelling taught philosophy at a number of prestigious German universities between 1798 and his death in 1854, including Jena, Wurzburg, Munich and, finally, Berlin, where he achieved something of the status of a prophet for the new movement. Although he himself made extensive use of contemporary physics and biology in his thought – and, indeed, regarded himself as providing the philosophical underpinnings for the natural sciences (to the point where some of his ideas would not be entirely out of place in modern quantum physics) – his *Naturphilosophie* gave a special prominance to art and religion. He emphasized the nature of the world as an organism (something not too remote from the Gaia hypothesis being advocated today). In his view, philosophy and poetry were two sides of the same coin, and he looked forward to the day when they would once again become one. He argued that we apprehend the true nature of the world directly rather than arriving at it by logical argument and empirical test. His views are graphically summed up by his observation that the seventeenth-century philosopher Leibnitz had asserted from intuition what the German Enlightenment philosopher Christian Wolff had later merely proved. To Schelling it was self-obvious whose was the greater contribution.

The Romantic movement's artistic contributions to the sum of human culture were, of course, enormous, especially in music. Its origins and psychological background are, however, deeply suspect. It smacks of a dubious anthropocentrism, an attempt to place humans at the centre of the universe merely because we experience that universe through our senses. It is redolent with a hankering after a purpose for life in a universe that contemporary astronomy and biology were inexorably demonstrating to be both accidental in origin and purposeless in existence. It was a conscious attempt to try and rectify what its supporters regarded as the awful wrong done to the human condition when Galileo unceremoniously displaced us from the centre of the cosmic stage. It was a desperate cry for help from a bewildered and frightened child who had just discovered that it was alone in a room that it had assumed was peopled with adults.

It has been argued that the extraordinary extent to which the Romantic movement permeated German culture and society during the nineteenth century, combined with Bismarck's *Kulturkampf* against the Catholics and the rise of nationalistic thinking in the philosophical writings of intellectuals like Nietzsche and the biologist Ernst Haeckel, prepared the ground for the rise of Nazism in the decades after Germany's crushing defeat in World War I. It would, of course, be naive to suggest that Romanticism was solely responsible for the rise of National Socialism, but its emphasis on nationalism and a directly experienced *Volksgeist* (or 'folk soul') was unquestionably an important source of inspiration. Moreover, there can be little doubt that the anti-scientific and anti-liberal views fostered by the Romantics contributed significantly to an intellectual climate that allowed Nazism to flourish unchallenged.

This is not to suggest that there is (or even should be) a stark dichotomy between science and the arts. It is simply wrong to suppose that science and culture are in opposition to each other, such that by opting for rational science we necessarily abandon the artistic and the emotional. The pursuit of science, as any active scientist will tell you, entails as much appreciation of the beauty of nature, of the elegance of ideas, as any artistic or literary endeavour. To see the exquisite beauty of a supernova through a large telescope, to watch the drama and tension of a wolf-pack at hunt, to find the delicate shades of a wild orchid tucked away in the crook of a tree root – these are the daily emotional fare of scientists the world over. They are the things scientists look forward to, that give them pleasure and sustain them through the dull and tedious business of normal science. It is that same sense of achievement one gets from placing the last piece in a giant jigsaw puzzle, or from finally managing to play a Paganini violin concerto without making a mistake – that same sense of satisfaction at completing a painting, or writing the last difficult line of a poem or preparing an exquisite dish. Here was a problem, something we did not understand; and now, by careful examination and the application of pure logic, combined with the experimental equivalent of green fingers, it has been possible to determine why the world *has* to be like that. The difference, perhaps, is that science is less narcissistic than art, preferring to marvel at the wonders of the external universe rather than becoming mesmerized by the fantasies of the human mind.

There is a deeper issue here that needs to be explored: it concerns the suggestion that scientific explanations are somehow incompatible with aesthetic experiences of the same phenomenon. It is difficult to understand why this view should be so widespread, but widespread it undoubtedly is. Take, for example, the purely culinary matter of cream and the rich taste it adds to both savoury and sweet foods. In what sense is our appreciation of this exquisite substance altered by the knowledge that cream is in fact a suspension of tiny spherical particles of fat in water? Cream whips (when plain milk won't) because the forces applied during vigorous whisking deform the fat particles into long thin ribbons that eventually interlink to produce a solid mass out of what had once been a liquid. But what conceivable relevance does that knowledge have to our subsequent gustatory experience of cream? The answer, of course, is none at all. We can marvel simultaneously at both the physical mechanism and the effect it has on our taste buds without having to feel that there is any conflict between these two experiences. The one does not diminish the other.

Indeed, understanding the physics and chemistry of culinary processes might well enable us to improve on what our ancestors achieved by simple trial and error. Knowing just why cream whips may allow us to design better machines to do the job. It may also allow us to reproduce the gustatory experience without the harmful side-effects of too much fat in our diet. In fact, the elastic properties of fat globules place a lower limit (at about 25 per cent) on the proportion of fat that cream must contain in order to whip. This knowledge is used in the production of 'whipping' creams: these have a lower fat content (about 30 per cent) than conventional 'double' cream straight from the cow (which is about 40 per cent fat).

The problem, of course, lies in a failure to recognize and distinguish between Tinbergen's Four Why's (see Chapter 5). The mere fact that we can offer an explanation at one level does not thereby automatically exclude the possibility that we can offer equally valid explanations at other levels. Cause *must* be separated from consequence.

I can do no better than to end this chapter by quoting the philosopher Dan Dennett's reflections on the way modern neurobiology has debunked pseudo-religious mystical views of human consciousness:

Looking on the bright side, let us remind ourselves of what has happened in the wake of earlier demystifications. We find no diminution of wonder: on the contrary, we find deeper beauties and more dazzling visions of the complexity of the universe than the protectors of mystery ever conceived. The 'magic' of earlier visions was, for the most part, a cover-up for frank failures of imagination, a boring dodge enshrined in the concept of a deus ex machina. Fiery gods driving golden chariots across the skies are simpleminded comic-book fare compared to the ravishing strangeness of contemporary cosmology, and the recursive intricacies of the reproductive machinery of DNA make élan vital about as interesting as Superman's dread kryptonite. When we understand consciousness – when there is no more mystery – consciousness will be different, but there will still be beauty, and more room than ever for awe.

(Dennett, p.25)

10
Divided Loyalties

To vote for relinquishing human power over nature through science is to vote for the permanent oppression of some people by others.

Andrew Collier: *Scientific Realism and Socialist Thought* (1989)

It was a kind of cloud that overshadowed knowledge for a while and blew over.

Francis Bacon: *De Augmentis Scientarum* (1606)

I have argued that much of the negative reaction towards science arises from a failure to understand just what it is that science is all about. Yet we are now all dependent on science to sustain our day-to-day lives. Neither we in the industrialized nations nor those in the developing world could return to pre-industrial agricultural economies without incurring terrible consequences in social and demographic terms.

Malthus may have been wrong about the demographic catastrophe that loomed over Europe in the early 1800s, but he was only wrong on a technicality. He could not have known how great would be the impact of the continuing agricultural and industrial revolutions that cascaded through the century following the publication in 1798 of his *Essay on the Principle of Population*; combined with the mass migrations to North America and Australasia, they managed to offset the alarming rate at which the British and other European populations were growing, and so forestalled the cataclysm that Malthus had so confidently forecast. Our current population levels are far in excess of those that could be sustained by a pre-industrial agricultural economy and, should we ever be forced to return to those conditions, Malthus's worst fears will surely be confirmed.

We are, to put it bluntly, now firmly locked into the scientific world whether we like it or not. We cannot afford to allow the future course of science to be diverted by parochial interests. It is incumbent on us to break the mould that produced the sense of divided loyalties that we have been heir to since Galileo. But that will not be easy, for, as I suggested in

Chapter 7, the human mind was not designed as a rational scientific mind. In a very real sense, we have to work against our natural instincts. There are no easy answers to offer at this time, but we must, and should, give the problem serious consideration before it is too late.

Two Cultures

When Galileo published his *Dialogue Concerning the Two Chief World Systems* in 1632, he was of course referring to the heliocentric and the geocentric models of the universe. But as he rapidly found to his cost, the debate went much deeper than trivial matters of astronomy. The old geocentric view of the universe underpinned (for no particularly good reason other than tradition) the tenets of the Christian Church. It stood for the old certainties, the emotional crutch that religions have provided for us since the dawn of human history. The heliocentric view with which he sought to replace it came to represent the brave new world of modern science, of man alone in the universe.

It would be easy to argue that much of the angst and hand-wringing that have been so conspicuous a feature of this debate owes its origin to our failure to come to terms with the Enlightenment's discovery that the human species is not, after all, the centrepiece of the universe, the reason for its existence. Were that all there was to it, the future would not be a matter of great concern. What we are witnessing now would simply be the last lingering doubts in the after-shock of Galileo's first astonished glimpse of the moon's surface through a telescope nearly four centuries ago. But the problem may not be so simple. It may have much more to do with the nature of being human. Emotion is so deep a part of our psychological make-up that it cannot be so easily rooted out.

Our problem is that we really know surprisingly little about the whole phenomenon. Is everyone really capable of absorbing the lessons of science or does science require special intellectual skills that only a privileged few are capable of acquiring? Are there, perhaps, people who are 'natural' scientists and others who are 'natural' artists? Or can we instil these skills in everyone, albeit by dint of long periods of training? On a more worrying note, is the rationality of science powerful enough to overcome the deep psychological need for faith and belief that so characterizes our species? If it is not, the advance of science may be creating a two-class polity in which there are those who are intellectually capable of learning about science and those who are not.

And if this really is so, then we may well be heading for the creation of a high priesthood that is beholden only to itself. Some would argue that we have already reached this point in certain areas of science (for example in government attempts to impose nuclear power or in scientists' own pursuit of areas like embryo research). This raises serious issues, for a disenfranchized majority will always be prone to fundamentalist proselytism in ways that will inevitably make the pursuit of science more difficult.

In the past few decades, literary theorists (operating under the influence of Postmodernist French social theorists like Derrida and Foucault) have repeatedly claimed that the methods of empirical science are not appropriate to many aspects of human cultural behaviour (a claim that has also been made vociferously by social anthropologists and others in the social sciences, including some archaeologists). Paradoxically, Postmodernism can be seen as a welcome advance in the direction of science on the part of the humanities, at least in the sense that it represents a realization of one of the processes fundamental to the scientific method, namely the challenging of existing assumptions and the generation of alternative hypotheses. Where Postmodernism fails its practitioners, however, is in its deliberate refusal to implement that other important cornerstone of science, the testing of hypotheses.

But the business of hypothesis-testing *is* something that we all need to use. Postmodernist claims notwithstanding, hypotheses about the real world are an inevitable part even of literary theory and they can – and *must* – be tested. It simply isn't good enough to present a series of alternative viewpoints without making any attempt to choose between them. There *are* standards of scholarship, and any proposition about the real world must be right or wrong. Even when that real world is a writer's mind-state or intentions when composing a work, it is perfectly possible, at least in principle, to test the validity of inferences and hypotheses. I do not doubt that such tests may be difficult to devise and sometimes less than perfect to implement. But that is nothing new in science. The way in which astronomers have used strong inference to explore the impossible on the galactic scale stands as a shining example of what can be done with the most unprepossessing of subject-matters. Similarly the new dimensions being opened up in the study of animal behaviour by recent advances in evolutionary theory demonstrate that not all of science needs to be reductionist, for here the emphasis is very much on the strategic aspects of what animals (and humans!) do.

In contrast, current Postmodernism seems to be little more than a cosy construction of reality intended to serve an essentially political end. It gives the impression of being nothing more than an intellectual bolt-hole for those whose own research programmes have achieved little over the past century and have even less to offer the modern world. To claim that the world is however we wish to interpret it is intellectual laziness and doesn't deserve the name of scholarship. We can, and should, do better than that. Science is not 'just another' vague theory, as some of the sociologists of knowledge would have us believe. The theories of science actually work (at least given time and patience!), and they work because the methods of science come as close as is possible in this imperfect world to guaranteeing success.

To be fair, of course, even scientists have not always been immune to such temptations. In his book *The Social Meaning of Modern Biology*, Howard Kaye takes the biologist Steven Rose and his colleagues to task for suggesting in their book *Not in Our Genes* that sociobiology is an attempt by the capitalist ruling clique to justify its continued domination of society. In so doing, he complains, they are mindlessly trotting out a party line that bears little or no relationship to reality. As Ulrica Segerstrale has pointed out, the self-styled Marxists Rose and Lewontin share with their arch-opponent, sociobiologist E. O. Wilson, a number of key objectives, including a commitment to political (as opposed to economic) liberalism, a desire to break the stranglehold that organized religion has over the minds and lives of ordinary men and women and a belief in the need for a planned and more equitable society. Their deep-rooted antagonism towards sociobiology is all the more baffling because many of the founding fathers of sociobiology (including, for example, the English biologists J. B. S. Haldane and John Maynard Smith) have impeccable socialist credentials.

The optimistic view is that these aberrations are a consequence of ignorance. It suggests that the apparent gulf between the sciences and the humanities may be just an attitude of mind; and, given that solutions to attitudes of mind are usually just matters of education, there is every reason to hope that progress can be made. But we cannot leave such things to fend for themselves. Science is a precarious edifice: it needs only a moment of neglect for it to crumble away to dust.

There are already disturbing signs that all is not well in this respect. In Britain, at least, there is an increasing tendency for the best students to specialize in the humanities rather than the sciences. From that experience, they emerge indoctrinated by teachers whose understanding of science is wholly inadequate and who tend to perpetuate their own views with conviction and persuasion. Nor is this a problem that is confined to the British Isles. Alan Cromer has argued forcefully in his book *Uncommon Sense* that the American educational system is in even worse shape in this respect. There, the Creationist lobby, with its fundamentalist religious beliefs, is a significant force that has been threatening to derail the entire process of scientific education for some time. To this has been added in recent years the contemporary versions of the Inquisition fostered by the 'politically correct' fringes of the humanities.

Whatever Happened to Renaissance Man?

What is clearly required is a much greater awareness of science, and that requires more and better science education. That ought not to conflict with an education in the humanities. But it might mean that we need a longer period at school, followed by a broader-based and longer period of higher education. Alan Cromer has, however, argued that, if science were taught in the right way (and he advocates a more practical hands-on approach), a sound education could be accomplished in *less* time than is currently given to it.

In the meantime, however, we urgently need more extensive popularization of science as well as a better quality of science journalism. Popular accounts of science must move away from the magical 'hero-quest' and 'gee-whiz' varieties to a more realistic attitude in which science is reported as it really is. If that requires changing people's reading habits, then it is a task we must address. This is a role that television is especially suited to play because of the immediacy of its impact. We urgently need more science on television.

Stephan Collini, in his essay introducing the recent reprint of Snow's *The Two Cultures*, has pointed to the importance of the contributions to the public understanding of science that have been made by popular books written by prominent scientists like Stephen Hawking, Richard Dawkins and Stephen Jay Gould. These books have a very different character to those written by journalists or non-specialist writers: they have an immediacy and sense of excitement that only someone intimately concerned with the business of science can hope to achieve. Professional scientists should be encouraged to write more for the popular market. Joseph Schwartz argues, in *The Creative Moment*, that there are encouraging signs that literature is beginning to take science seriously as a topic to explore. He points to Primo Levi's use of industrial chemistry in this way.

But if we are to achieve any long-term success, we will need to look more carefully at science teaching. Somehow we are failing to fire young children's enthusiasm. The sciences are perceived as being harder than the humanities because the answer is either right or wrong and more obviously factual, whereas in the humanities you can get away with waffle even if you don't know what the answer really should be. Right or wrong, this common perception will not be easy to overcome, for there is a natural asymmetry between the arts and sciences in this respect. In the sciences you need an information pack before you can start to discuss anything, and those tools may have to be acquired by rote learning before we can do something with them. Not everyone intuitively understands the principles of photosynthesis, but everyone knows what a good story is or how to sing a song. Anyone can paint (even if the results are hardly in Michelangelo's class), but not everyone can write out the geometric proof of Pythagoras's theorem (as I recently discovered when trying to do it for my son).

We are caught in a cleft stick here. Can we instil the basic tools that children need to do science at an early painless stage, just as we teach them to read? Or is science more like learning to play a musical instrument – hours of enforced agony that only begin to pay their dividends of pleasure years later, and then only for the gifted few? If the scientific equivalent of the tone deaf (those who could no more construct a scientific argument or do an experiment than play a musical instrument) really does exist, then our task may be particularly difficult. Yet I am encouraged by the fact that, speaking for myself, even the tone deaf can learn to appreciate music.

Alan Cromer has pleaded for a more innovative approach based on learning the principles of science and mathematics through the use of concrete examples. This is something he has tried to put into practice in a small number of schools within the USA. I am not entirely convinced that this will remedy the situation, because there is a perennial danger that the kinds of exercises that can be done with young children will often seem to be pointless and trivial. It might be that there are simply no real alternatives to rote learning: tennis stars, after all, are not made overnight, but by hours upon hours of tediously repetitious practice. But I do agree with him that the root of the problem lies in the fact that science teachers themselves are not trained in the right way. Teaching science requires boundless enthusiasm for the subject and that in turn requires confidence in one's own ability to answer questions and fire up young people's imaginations. Cromer has tried using retired scientists as specialist teachers whose sole task is to arouse children's interest in science, and he claims great success for the project, at least at the local level. Pupils gain immeasurably from working with people who genuinely know what they are doing, and the scientists concerned gain pleasure from a renewed sense of purpose, from the opportunity to give back to society a little of what they themselves gained in their own education. The drawbacks, of course, lie in the problems of presenting technical subjects at a level that the layman can understand (see Chapter 8). Such problems are obviously exacerbated when Nobel Prize-winners in particle physics try to talk to eight-year-old children.

There are other possibilities, of course. One would be to lengthen degree courses by an extra year, perhaps by adding what would in effect be a masters year onto the end. In Britain, we are already beginning to see this happening. In physics, for example, most British universities now run a two-tier system. Undergraduates take the conventional three-year BSc but emerge from the process with an Ordinary degree. If they want to do an Honours degree (the normal degree awarded by UK universities), they must take an additional honours year.

The reason for this is simply that students enrolling in university physics courses no longer have sufficient high school math to cope with the mathematical requirements of modern physics. They need the extra year to bring them up to speed. There is some evidence of a similar problem emerging in the United States. The National Science Foundation's 1993 survey Science and Engineering Indicators noted that, over the preceeding fifteen years, around 20% of all freshmen who intended to major in science or engineering felt that they needed remedial work in math.

An alternative might be to expect everyone leaving school to undertake a form of science national service. This could involve a period spent working as a research assistant in a laboratory or in industry, or taking part in a field study or a conservation programme. Those leaving school at sixteen might be involved in something more practically oriented, while those leaving school at eighteen with a view to going on to higher education might be involved in something more intellectually challenging at the frontiers of science or technology. The benefits of real-world hands-on experience would be of great advantage in itself, but the additional year would result in more mature university graduates which would, in my experience, be a significant advantage in its own right. If we allowed those intent on leaving school at sixteen to spend the whole of the last year of their education engaged in this kind of programme, perhaps it would alleviate some of the problems of boredom and truancy currently causing so much concern in this age group.

The cost would, of course, not be negligible. But we have to ask not just about the cost to the taxpayer, but also about the benefits to be gained in the future by society as a whole. If we want to pass on a secure economy to our children, then this may be money that we can ill afford to quibble over.

A more radical solution would involve the complete restructuring of the way we teach science. It is clear that young people find the physical sciences harder to come to grips with than the life sciences. This being so, we should consider focusing school science on what has dismissively been termed the 'soft' sciences, namely animal behaviour, ecology and psychology. These are areas of science that are intuitively easier to understand because we can relate to their subject-matter more easily. Using these more amenable disciplines, it would be possible to introduce the principles of hypothesis-testing and the methods of science, as well as concepts like Tinbergen's Four Whys (with its emphasis on tackling problems at several different equally valid levels). The advantage of this lies precisely in the fact that these disciplines deal predominantly with social behaviour, and would thus capitalize on children's natural facility for understanding the social world.

These disciplines could constitute the entire corpus of science taught right through the secondary school curriculum. But since these areas naturally lead into mathematics and the physical sciences (via the chemical bases of genetics and the physiological processes that underpin animal behaviour), there could be a natural progression on to the physical sciences in post-sixteen education. Students would be led into chemistry by asking what kinds of mechanistic processes underpin an animal's behaviour. This would lead naturally into a discussion of energetics, for example, which would in turn raise questions about the chemistry of the nutritional constituents of food. This would allow students to appreciate the natural hierarchy of the sciences and the way they form an integrated structure – something that they cannot do at present, given the way the sciences are taught.

One consequence, of course, is that students entering higher education would inevitably have a very much more limited grounding in physics and chemistry, and perhaps even mathematics. Academics in the physical sciences may simply have to bite the bullet on that one in the interests of better-quality students in the long run. It would, of course, necessitate an extended period of university education in the physical and mathematical sciences. Governments of all persuasions will certainly baulk at the cost, but I am not sure we really have much choice in the long term. The situation is already so parlous in terms of the quality of applicants to university science courses that we might just as well give science places to students whose primary qualifications are wholly confined to the humanities.

I am only too conscious of the fact that physical scientists will look aghast at this last suggestion. But can the prospects of taking on someone who is a high flier in the humanities be any worse than taking on someone with bottom-of-the-barrel grades in the sciences? Would physics courses be better off, for example, taking on undergraduate students who already had BA degrees in English Literature than students fresh out of high school with minimal qualifications in the sciences? After all, the fact that someone has a top-flight grade point average in, say, history, does at least tell you that he has the ability and persistence to master a subject. With a little extra coaching in math and some commitment and support, he would surely do better in a science course than an unmotivated weaker science student.

I can only speak for myself in this respect. My senior high school qualifications were confined exclusively to the humanities: I did no science of any kind after the age of 16. But that does not seem to have held me back when it has come to the more mathematical and chemical aspects of those areas where I now do my research and teaching (animal behaviour and ecology, both of which are now very mathematical). I learned my science and my math piecemeal over an extended period of time because my interests and research motivated me to find out about them.

If this is a plausible solution, then the answer may be to offer humanities graduates junior fellowships that would allow them to spend the next three years on intensive science courses. At root, it is simply a question of attitude: we need only to persuade students from the humanities that science offers exciting and worthwhile careers and that a humanities background is no bar to them.

The physicist Dick Feynman firmly believed that science ought to be fun, and so it should be. That it can be entertaining is reflected in the number of Discover Science exhibitions that have sprung up over the past decade. At these exhibitions children (and adults!) can engage in 'hands-on' science. Yet entertaining as these places are, they focus predominantly on the 'gee-whiz' aspects of science. They function at best simply to stimulate people to find out more. Ultimately everything will hinge on training good-quality science teachers who can inspire young people to go beyond the surface glitz to the exciting world of ideas. History is no bad way into this, so there probably ought to be more place in the science curriculum for the philosophy and history of science. This would enable students to get a better feel for what science is all about and how it grew from small beginnings into the monumental edifice of today.

One thing is certain: if we want to survive the next century or so, we need to change radically the way we teach science. Such changes will take time to implement, and their impact on the public domain will take even longer – possibly in the order of a generation or more before their real benefit is felt. But time is not on our side, for we may be destroying the planet faster than we realize. Survival is both a matter of the will to survive and a race against a clock that we inadvertently set ticking a quarter of a million years ago when we first evolved the brain that subsequently allowed us to invent religion and do science. The choice is ours, but it is our children who will pay the price for the wrong decision.

A Salutary Lesson

Let me end with a cautionary tale. It concerns the unfortunate Vikings who colonized Greenland during the early Middle Ages and whose colony died out some time during the first few decades of the fifteenth century. It serves to remind us of what can happen when a society clings to its old views and fails to adapt its behaviour to the shifting realities of the world in which all of us, scientist and artist alike, are perforce obliged to live.

When the Vikings first settled the ice-free southern coast of Greenland during the closing decades of the tenth century, they brought with them the full panoply of the culture and farming practices that had served them so well in Scandinavia. The island's discoverer, Eric the Red, is said to have named it Greenland despite the enormity of its ice cap in order to encourage others to join him in colonizing the place – perhaps the earliest recorded example of questionable real estate advertising. Yet, despite this unpromising start, the Viking settlers initially met with considerable success. Their farms prospered and their cattle herds grew. At its peak the colony amounted to some 3000 souls scattered in 280 farmsteads around the southwestern coast. By the twelfth century they even had their own bishop and their own parliament.

But as temperatures in the northern hemisphere began to decline with the onset of the Little Ice Age of the mid-fourteenth century, things began to go dreadfully wrong. The cattle were unable to survive the harsher winters and the shorter growing season reduced the already rather poor corn crops so dramatically that starvation set in. This much is clear from the increased frequencies with which rickets and short stature (both signs of malnutrition) are found in the skeletons from later burials. Finally, some time after 1410 when the last supply ship sailed back to its home port in Norway, the Greenland community died out.

It seems that the problem lay mainly in the community's refusal to adapt its cultural practices to suit the new conditions. The Vikings came with a technology and lifestyle that had evolved to meet the conditions of the Scandinavian peninsula in the final quarter of the first millennium AD. But Scandinavia differs from Greenland in one crucial respect: it lies in the Gulf Stream which brings warm water from the Caribbean and its climate is much milder in consequence. Although Greenland's southernmost point, Cape Farewell, lies at the same latitude as Oslo in Norway, and Godthaab (where the most northerly of the Greenland settlements were concentrated) is no further north than Trondheim, the climate at these Greenland sites is far more severe, even today.

T. H. McGovern has suggested that the Vikings' principal problem lay in their refusal to learn from the Eskimos with whom they were just beginning to come into contact. Unlike the seventeenth-century English immigrants to North America who adopted the agricultural techniques and hunting practices of their Indian hosts, the Vikings declined to benefit from the Eskimos' long experience of how to survive in the Arctic. Although they eventually took to hunting seals and walruses at their breeding grounds on the sea ice to the north of their settlements, the Vikings continued to use the inappropriate hunting techniques developed in Scandinavia by their forebears rather than those of the Eskimos.

There are several reasons why the Vikings might have been so recalcitrant. The one favoured by McGovern is that their cultural background was too inflexible and that they viewed the pagan Eskimos as uncivilized. But such an explanation begs questions as to why the Viking culture was so inflexible when the Eskimos' culture was not. It also assumes an unnecessarily pessimistic view of people's abilities to recognize cause–effect relationships in the natural world, as well as implying a truly monolithic view of culture. Neither seems especially appropriate given the evidence for the often fine-tuned adaptations shown by human cultural practices the world over (see Chapter 3).

It seems more plausible that the Vikings had the bad luck to find themselves at the point of convergence of at least three separate factors, each of which played an important contributory role in their final demise, even though individually none of them would have proved fatal on its own. One of these has to be the fact that they had survived successfully in Greenland for the better part of three centuries using their Scandinavian technology. Unlike Captain Smith's early Virginians, for whom evidence of their inability to cope was forced upon them the very first winter, the Vikings had three centuries of evidence to suggest that their technology was appropriate for survival in Greenland. When cultural practices are known to work, there is always considerable resistance to change (as has been shown by the elegant analyses of cultural evolution undertaken by Robert Boyd and Peter Richerson). Like Tony Dickinson's rats, the Vikings had learned their habits of thought too well and found it difficult to step back far enough from the circumstances of the moment to see that adjustment was necessary.

Secondly, the climatic changes crept up on the Vikings just a little too slowly. They had experienced bad years before, but things had always improved in the long run. By the time they came to recognize that the increasing hardship reflected a genuine linear trend rather than part of the normal year-to-year variation, it was too late to do much about it. That this is probably what happened is suggested by the evidence for a gradual change of diet during this later period: it seems that the Vikings did eventually begin to make increasing use of the sea mammals (by far the most abundant and accessible source of food). But they lacked the skills that the Eskimos had built up over centuries, and so proved to be inefficient hunters.

The third, and perhaps most important, factor must have been that the Vikings only began to make contact with the Eskimos right at the end of the crucial period. The Eskimos had traditionally occupied the northern Arctic parts of Greenland, and had only begun to migrate southwards as the temperatures declined in the fourteenth century, at which point they first came across the Viking expeditions exploring the sealing sites around Disko Bay, well to the north of the Viking settlements. With little direct evidence to suggest that Eskimo survival skills were any better than their own, there would have been no incentive for the Vikings to copy the Eskimos' hunting techniques. In any case, survival would ultimately have depended on learning the Eskimos' transhumant, navigational and igloo-building techniques, not just their hunting skills, for the Vikings' stone-built houses left them with insufficient flexibility to move in the wake of the migrating sea mammal herds.

In short, the changes that were required to ensure the Vikings' survival on the Greenland ice cap were too great in too short a time-scale. To be sure, the Vikings' arrogant attitudes towards the pagan Eskimos no doubt made them less willing to learn from the latter's greater experience, but this can be only one among a number of contributory factors. Even arrogance will eventually bow its head in the face of conspicuous failure, given time to learn, as the fate of the early Virginia colonies shows.

Why, then, did the Eskimos succeed where the Vikings failed? The crucial point is that they had longer to learn. The Eskimos first seem to have crossed the Bering Straight into North America from their ancestral homelands in northeastern Asia some 12,000 years ago at the tail-end of the last Ice Age. By then they had, of course, built up considerable experience of living under Arctic conditions. Yet these skills were surely not acquired without cost: buried beneath the tundras of the Siberian peninsula must be the remains of more than one Eskimo version of the Greenland Viking community.

One other factor must surely have been important in the demise of the Vikings: Greenland is an island. The ancestral Eskimos of Siberia could always retreat southwards to remain within their preferred habitat as the icefields expanded, thereby buying time in which to acquire the skills needed to survive in the Arctic. But the Greenland Vikings could not: occupying the southernmost shores of the island, they had nowhere to retreat to as the icefields moved relentlessly southwards. They could only stay put and do their best.

The Greenland Vikings were neither the first nor the last society to fail to notice that changing times required new approaches to the business of survival. The archaeological sites of the world, from the ruins of the Mesopotamian city-states to the Aztec temples of Mexico, testify to the fact that the lessons of nature are sometimes learned too slowly. Yet if the fate of the Vikings stands as mute testimony to the terrible consequences that await those who fail to challenge their assumptions about how the world works, we should not be unduly pessimistic. The success of the Greenland Eskimos in coping with precisely the same conditions should remind us that survival is possible when change is guided by careful attention to learning the lessons of past successes and failures.

Bibliography

Chapter 1

Anon (1985). 'What do people think of science?', *New Scientist* (21 Feb): 12–16.

Appleyard, B. (1992). *Understanding the Present: Science and the Soul of Modern Man*. Pan, London.

Best, S. and Kellner, D. (1991). *Postmodern Theory: Critical Interrogations*. Macmillan, London.

Highfield, R. (1992). 'A small tonic for everyone in the lab', *Daily Telegraph* (24 Aug): 11.

Hugill, B. (1994). 'Cultists go round in circles', *Observer* (28 Aug): 3.

Kenny, M. (1991). 'Of mice and Mengele', *Evening Standard* (9 May).

Martin, D. (1994). 'Don't put chemistry on the back burner', *Observer* (review supplement) (11 Sept): 12.

Martin, B., Irvine, J. and Turner, R. (1984). 'The writing on the wall for British science', *New Scientist* (8 Nov): 25–29.

Ouspensky, P. D. (1939). *Tertium Organum*. Kegan Paul, Trench Trubner, London. (Russian original 1921)

Snow, C. P. (1993). *The Two Cultures*. Cambridge University Press, Cambridge.

Chapter 2

Bacon, F. (1620 [1889]). *Novum Organon*, trans. T. Fowler. Oxford University Press, Oxford.

Bowler, P. J. (1986). *The Idea of Evolution*. University of California Press, Los Angeles.

Chalmers, A. F. (1978). *What is This Thing Called Science?* Open University Press, Milton Keynes.

Coplestone, F. (1950). *A History of Philosophy*, vol. 2: *Mediaeval Philosophy*. Burns Oates, London.

Coplestone, F. (1953). *A History of Philosophy*, vol. 3: *Ockham to Suarez*. Burns Oates, London.

Coplestone, F. (1959). *A History of Philosophy*, vol. 5: *The British Philosophers*. Burns Oates, London.

Dunbar, R. I. M. (1982). 'Structure of social relationships in a captive group of gelada baboons: a test of some hypotheses derived from a wild population', *Primates* 23: 89–94.

Dunbar, R. I. M. (1988). *Primate Social Systems.* Chapman & Hall, London.

Feyerabend, P. (1975). *Against Method: Outline of an Anarchistic Theory of Knowledge.* New Left Books, London.

Gale, G. (1979). *Theory of Science.* McGraw-Hill, Toronto.

Hempel, C. G. (1966). *Philosophy of Natural Science.* Prentice Hall, Englewood Cliffs, NJ.

Kuhn, T. S. (1970). *The Structure of Scientific Revolutions,* 2nd edn. University of Chicago Press, Chicago.

Lakatos, I. (1978). *The Methodology of Scientific Research Programmes,* vol. 1. Cambridge University Press, Cambridge.

Lossee, J. (1972). *A Historical Introduction to the History of Science.* Oxford University Press, Oxford.

Mayr, E. (1982). *The Growth of Biological Thought.* Belknap Press, Cambridge, MA.

O'Hear, A. (1985). 'Popper and the philosophy of science', *New Scientist* (22 Aug): 43–45.

Popper, K. R. (1959). *The Logic of Scientific Discovery.* Hutchinson, London.

Quine, W. O. (1961). *From a Logical Point of View.* Harper and Row, New York.

Ryle, G. (1949). *The Concept of Mind.* Hutchinson, London.

Chapter 3

Aristotle (1943). *Generation of Animals,* ed. A. L. Peck. Heinemann, London.

Aristotle (1965). *Historia Animalium,* vols 1–2, ed. A. L. Peck. Heinemann, London.

Aristotle (1972). *Parts of Animals,* vols 1–3, trans. D. M. Balme. Heinemann, London.

Atran, S. (1990). *Cognitive Foundations of Natural History: Towards an Anthropology of Science.* Cambridge University Press, Cambridge.

Aubert, G. and Newsky, B. (1949). 'Note on the vernacular names of the soils of the Sudan and Senegal', *Proceedings of the First Commonwealth Conference on Tropical and Subtropical Soils.* Commonwealth Bureau of Soil Sciences, Technical Communication no. 46. Harpenden.

Barrow, J. (1988). *The World Within the World.* Oxford University Press, Oxford.

Diamond, J. (1966). 'Zoological classification system of a primitive people', *Science* 151: 1102–1104.

Dunbar, R. I. M. (1991). 'Foraging for nature's balanced diet', *New Scientist* 131: 25–28.

Dunbar, R. I. M. (1993). 'Seeing biology through Aristotle's eyes', *New Scientist* 137: 39–42.

Durkheim, E. (1915, 1982). *Elementary Forms of Religion.* Allen & Unwin, London.

Durkheim, E. (1924, 1953). *Sociology and Philosophy.* Cohen & West, London.

Finch, V. and Western, D. (1977). 'Cattle colors in pastoral herds: natural selection or social preference', *Ecology* 58: 1384–1392.

Hacking, I. (1983). *Representing and Intervening.* Cambridge University Press, Cambridge.

Horton, R. (1993). *Patterns of Thought in Africa and the West: Essays on Magic, Religion and Science.* Cambridge University Press, Cambridge.

Huff, T. (1993). *The Rise of Early Modern Science.* Cambridge University Press, Cambridge.

Isack, H. A. and Reyer, H.-U. (1989). 'Honeyguides and honey gatherers: interspecific communication in a symbiotic relationship', *Science* 243: 1343–1346.

Joseph, G. C. (1990). *The Crest of the Peacock.* Penguin, Harmondsworth.

Kirch, P. (1984). *The Evolution of Polynesian Chiefdoms.* Cambridge University Press, Cambridge.

Knight, C. (1990). *Blood Relations.* Yale University Press, New Haven.

Leach, E. (1954). *Political Systems of Highland Burma.* London.

Lévi-Strauss, C. (1966). *The Savage Mind.* Weidenfeld & Nicholson, London.

Lewis, D. (1978). *The Voyaging Stars.* Collins, Sydney.

Lloyd, G. E. R. (1991). *Methods and Problems in Greek Science.* Cambridge University Press, Cambridge.

Lott, D. and Hart, B. L. (1979). 'Applied ethology in a nomadic cattle culture', *Applied Animal Ethology* 5: 309–319.

Malinowski, B. (1935). *Coral Gardens and their Magic.* Allen & Unwin, London.

Nakayama, S. and Sivin, N. (eds) (1973). *Chinese Science: Explorations of an Ancient Tradition.* MIT Press, Cambridge, MA.

Niamir, M. (1990). *Herders' Decision-making in Natural Resources Management in Arid and Semi-arid Africa.* FAO, Rome.

O'Dea, K. (1992). 'Traditional diet and food preferences of Australian Aboriginal hunter-gatherers', *Foraging Strategies and Natural Diet of Monkeys, Apes and Humans*, ed. A. Whiten and E. M. Widdowson: 73–82. Oxford University Press, Oxford.

Richards, P. (1989). 'Farmers also experiment: a neglected intellectual resource in African science', *Discovery and Innovation* 1: 19–24.

Richards, P. (1992). 'Rural development and local knowledge: the case of rice in central Sierra Leone', *Entwicklungs-ethnologie* 1: 33–42.

Ronan, C. A. and Needham, T. (1978, 1981). *The Shorter Science and Civilization in China.* Vols. 1 and 2. Cambridge University Press, Cambridge.

Sahlins, M. (1972). *Stone Age Economics.* Aldine, Chicago.

Skorupski, J. (1976). *Symbol and Theory.* Cambridge University Press, Cambridge.

Speth, J. D. (1992). 'Protein selection and avoidance strategies of contemporary and ancestral foragers', *Foraging Strategies and Natural Diet of Monkeys, Apes and Humans*, ed. A. Whiten and E. M. Widdowson: 105–110. Oxford University Press, Oxford.

Stroup, E. D. (1985). 'Navigating without instruments: the voyage of Hokule'a', *Oceanus* 28: 68–75.

Tambiah, S. J. (1990). *Magic, Science, Religion and the Scope of Rationality.* Cambridge University Press, Cambridge.

Western, D. and Dunne, T. (1979). 'Environmental aspects of settlement site decisions among pastoral Maasai', *Human Ecology* 7: 75–93.

Wilson, K. B. (1990). *Ecological Dynamics and Human Welfare: A Case Study of Population, Health and Nutrition in Southern Zimbabwe.* PhD thesis, University of London.

Chapter 4

Astington, J. W. (1994). *The Child's Discovery of the Mind.* Fontana Press, London.

Berlin, B. and Kaye, P. (1969). *Basic Color Terms: Their Universality and Evolution.* University of California Press, Berkeley.

Cheney, D. L. and Seyfarth, R. M. (1990). *How Monkeys See the World.* Chicago University Press, Chicago.

Clark, E. V. (1982). 'The young word maker: a case study of innovation in the child's lexicon', *Language Acquisition*, ed. E. Wanner and L. R. Gleitman: 390–425. Cambridge University Press, Cambridge.

Dasser, V. (1988). 'Mapping social concepts in monkeys', *Machiavellian Intelligence*, ed. R. Byrne and A. Whiten: 85–93. Oxford University Press, Oxford.

Dickinson, A. (1980). *Contemporary Animal Learning Theory.* Cambridge University Press, Cambridge.

Dickinson, A. (1985). 'Actions and habits: the development of behavioural autonomy', *Philosophical Transactions of the Royal Society* (*London*), B, 308: 67–78.

Domjam, M. and Wilson, N. E. (1972). 'Specificity of cue to consequences in aversion learning in the rat', *Psychonomic Science* 26: 143–145.

Dunbar, R. I. M. (1984). 'Ever since Descartes', *New Scientist* 103 (16 Aug): 32–34.

Dunbar R. I. M. (1989). 'Common ground for thought', *New Scientist* 121 (7 Jan): 48–50.

Harris, P. L. (1991). 'The work of imagination', *Natural Theories of Mind*, ed. A. Whiten: 283–304. Blackwell, Oxford.

Herrnstein, R. D. (1985). 'Riddles of natural classification', *Philosophical Transactions of the Royal Society* (*London*), B, 308: 129–143.

Holland, P. C. and Straub, J. J. (1979). 'Differential effects of two ways of devaluing the unconditioned stimulus after Pavlovian appetitive conditioning', *Journal of Experimental Psychology: Animal Behaviour Processes* 5: 65–78.

Huffman, M. A. and Seifu, M. (1989). 'Observations on the illness and consumption of a possibly medicinal plant, *Vernonia amygdalina* (Del.), by a wild chimpanzee in the Mahale Mountains National Park, Tanzania', *Primates* 30: 51–63.

Huffman, M. A. and others (1993). 'Further observations on the use of the medicinal plant, *Vernonia amygdalina* (Del), by a wild chimpanzee, its possible effect on parasite load, and its phytochemistry', *African Study Monographs* 14: 227–240.

Isack, H. A. and Reyer, H.-U. (1989). 'Honeyguides and honey gatherers: interspecific communication in a symbiotic relationship', *Science* 243: 1343–1346.

Jitsumori, M. and Matsuzawa, T. (1991). 'Picture perception in monkeys and pigeons: transfer of rightside-up versus upside-down discrimination of photographic objects across conceptual catgories', *Primates* 32: 473–482.

Leslie, A. M. (1982). 'The perception of causality in infants', *Perception* 11: 173–186.

Leslie, A. M. and Keeble, S. (1987). 'Do six-month-old infants perceive causality?', *Cognition* 25: 265–288.

Matsuzawa, T. (1985). 'Colour naming and classification in a chimpanzee (*Pan troglodytes*)', *Journal of Human Evolution* 14: 283–291.

Michotte, A. (1963). *The Perception of Causality.* Basic Books, New York.

Newton, P. N. and Nishida, T. (1990). 'Possible buccal administration of herbal drugs by wild chimpanzees, *Pan troglodytes*', *Animal Behaviour* 39: 798–799.

Premack, D. and Premack, A. J. (1983). *The Mind of an Ape.* Norton, New York.

Robertson, S. S. and Suci, G. J. (1980). 'Event perception by children in the early stages of language production', *Child Development* 51: 89–96.

Chapter 5

Bennett, J. (1991). 'How to read minds in behaviour: a suggestion from a philosopher', *Natural Theories of Mind*, ed. A. Whiten: 97–108. Blackwell, Oxford.

Hacking, I. (1983). *Representing and Intervening.* Cambridge University Press, Cambridge.

Horton, R. (1993). *Patterns of Thought in Africa and the West: Essays on Magic, Religion and Science.* Cambridge University Press, Cambridge.

Howson, C. and Urbach, P. (1989). *Scientific Reasoning: The Bayesian Approach.* Open Court, La Salle, IL.

Huxley, J. S. (1942). *Evolution: The Modern Synthesis.* Allen & Unwin, London.

Johnson-Laird, P. (1982). *Mental Models.* Cambridge University Press, Cambridge.

Maxwell, N. (1974). 'The rationality of scientific discovery, Part II. An aim-oriented theory of scientific discovery', *Philosophy of Science* 41: 245–295.

Newton-Smith, W. H. (1981). *The Rationality of Science.* Routledge & Kegan Paul, London.

Popper, K. R. (1979). *Objective Knowledge*, rev. edn. Oxford University Press, Oxford.

Rescher, N. (1980). *Induction*. Blackwell, Oxford.

Salmon, W. C. (1966). *The Foundations of Scientific Inference.* University of Pittsburg Press, Pittsburg, PA.

Scriven, M. (1959). 'Explanation and prediction in evolutionary theory', *Science* 130: 477–482.

Smith, P. K. (1981). *Realism and the Progress of Science.* Cambridge University Press, Cambridge.

Tinbergen, N. (1963). 'On the aims and methods of ethology', *Zeitschrift für Tierpsychologie* 20: 410–433.

Chapter 6

Avery, O. T., Macleod, C. M. and McCarty, M. (1944). 'Induction of transformation by a deoxyribonucleic acid fraction isolated from pneumococcus type III', *Journal of Experimental Medicine* 79: 137–158.

Barrow, J. D. (1988). *The World Within the World.* Oxford University Press, Oxford.

Coplestone, F. (1975). *History of Philosophy*, vol. 9: *Maine de Biran to Sartre.* Search Press, London.

Gale, G. (1979). *Theory of Science.* McGraw-Hill, Toronto.

Hamilton, W. D. (1964). 'A genetical theory of behavioural evolution: I, II', *Journal of Theoretical Biology* 7: 1–52.

Harrison, E. R. (1981). *Cosmology: The Science of the Universe.* Cambridge University Press, Cambridge.

Hull, D. (1974). *Philosophy of Biological Science.* Prentice Hall, Englewood Cliffs, NJ.

Huxley, J. S. (1942). *Evolution: The New Synthesis.* Allen & Unwin, London.

Mayr, E. (1982). *The Growth of Biological Thought.* Belknap Press, Cambridge, MA.

Platt, J. R. (1964). 'Strong inference', *Science* 146: 347–353.

Tversky, A. and Kahneman, D. (1981). 'The framing of decisions and the psychology of choice', *Science* 211: 453–458.

Tversky, A. and Kahneman, D. (1983). 'Extensional versus intuitive reasoning: the conjunction fallacy in probability judgement', *Psychological Review* 90: 293–315.

Wason, P. C. (1966). 'Reasoning', *New Horizons in Psychology*, ed. B. M. Foss: 135–151.

Weinberg, S. (1977). *The First Three Minutes: A Modern View of the Origin of the Universe*. Basic Books, New York.

Wolpert, L. (1992). *The Unnatural Nature of Science*. Faber and Faber, London.

Ziman, J. (1978). *Reliable Knowledge: An Exploration of the Grounds for Belief in Science*. Cambridge University Press, Cambridge.

Chapter 7

Astington, J. W. (1994). *The Child's Discovery of the Mind*. Fontana Press, London.

Avis, J. and Harris, P. L. (1991). 'Belief–desire reasoning among Baka children: evidence for a universal conception of mind', *Child Development* 62: 460–467.

Axelrod, R. (1984). *The Evolution of Cooperation*. Penguin Books, Harmondsworth.

Axelrod, R. and Hamilton, W. D. (1981). 'The evolution of cooperation', *Science* 211: 1390–1396.

Barkow, J. H., Cosmides, L. and Tooby, J. (eds) (1993). *The Adapted Mind*. Oxford University Press, Oxford.

Blurton-Jones, N. (1991). 'Tolerated theft: suggestions about the ecology and evolution of sharing, hoarding and scrounging', *Primate Politics*, ed. G. Schubert and R. D. Masters: 170–187. Southern Illinois University Press, Carbondale, IL.

Byrne, R. and Whiten, A. (eds) (1988). *Machiavellian Intelligence*. Oxford University Press, Oxford.

Cheney, D. L. (1977). 'The acquisition of rank and the development of reciprocal alliances among free-ranging immature baboons', *Behavioural Ecology and Sociobiology* 2: 203–218.

Cheney, D. L. and Seyfarth, R. M. (1990). *How Monkeys See the World*. Chicago University Press, Chicago.

Cosmides, L. and Tooby, J. (1993). 'Cognitive adaptations for social exchange', *The Adapted Mind: Evolutionary Psychology and the Generation of Culture*, ed. J. H. Barkow, L. Cosmides and J. Tooby: 162–228. Oxford University Press, Oxford.

Datta, S. (1983). 'Relative power and the maintenance of dominance', *Primate Social Relationships*, ed. R. A. Hinde: 103–112. Blackwell Scientific, Oxford.

Dunbar, R. I. M. (1988). *Primate Social Systems*. Chapman & Hall, London.

Dunbar, R. I. M. (1992). 'Neocortex size as a constraint on group size in primates', *Journal of Human Evolution* 20: 469–493.

Enquist, M. and Leimar, O. (1993). 'The evolution of cooperation in mobile organisms', *Animal Behaviour* 45: 747–757.

Flavell, J. H. and others (1983). 'A comparison between the development of the appearance–reality distinction in the People's Republic of China and the United States', *Cognitive Psychology* 15: 459–466.

Gardner, D. and others (1988). 'Japanese children's understanding of the distinction between real and apparent emotion', *International Journal of Behavioural Development* 11: 203–218.

Hardin, G. (1968). 'The Tragedy of the Commons'. *Science* 162: 1243–1248.

Hardin, G. (1977). *The Limits of Altruism*. Indiana University Press, Bloomington.

Keverne, E. B. (1982). 'Olfaction and the reproductive behaviour of nonhuman primates', *Primate Communication*, ed. C. T. Snowdon, C. H. Brown and M. R. Petersen: 396–412. Cambridge University Press, Cambridge.

Kummer, H. (1982). 'Social knowledge in free-ranging primates', *Animal Mind – Human Mind*, ed. D. R. Griffin: 113–130. Springer, Berlin.

Leslie, A. M. (1987). 'Pretence and representation in infancy: the origin of "Theory of Mind" ', *Psychological Review* 94: 412–426.

Menzel, E. (1974). 'A group of young chimpanzees in a 1-acre field: leadership and communication', *Behaviour of Nonhuman Primates*, vol. 5, ed. A. M. Schrier and F. Stollnitz: 83–153. Academic Press, New York.

Nishida, T. (1983). 'Alpha status and agonistic alliance in wild chimpanzees (*Pan troglodytes schweinfurthii*)', *Primates* 24: 318–336.

Waal, F. de (1982). *Chimpanzee Politics*. Unwin, London.

Waal, F. de (199.). *Primate Politics*.

Whiten, A. (ed.) (1991). *Natural Theories of Mind*. Blackwell, Oxford.

Whiten, A. and Byrne, R. (1988). 'Tactical deception in primates', *Behavioural and Brain Sciences* 11: 233–273.

Chapter 8

Appleyard, B. (1992). *Understanding the Present: Science and the Soul of Modern Man*. Pan, London.

Bolter, J. D. (1984). *Turing's Man: Western Culture in the Computer Age*. Duckworth, London.

Davies, P. C. W. (1979). *The Forces of Nature*. Cambridge University Press, Cambridge.

Dawkins, R. (1976). *The Extended Phenotype*. Freeman, San Francisco.

Dawkins, R. (1976). *The Selfish Gene*. Oxford University Press, Oxford.

Feynman, R. P. (1985). *Surely You're Joking, Mr Feynman!* Unwin, London.

Hamilton, W. D. (1964). 'The genetical evolution of social behaviour, I, II, *Journal of Theoretical Biology* 7: 1–52.

Haraway, D. (1989). *Primate Visions*. Routledge, New York.

Hawking, S. W. (1988). *A Brief History of Time*. Bantam, London.

Ingelfinger, F. J. (1973). 'The medical article revised', *New England Journal of Medicine* 289: 268–269.

Kaye, H. (1986). *The Social Meaning of Modern Biology.* Yale University Press, New Haven, CT.

Koestler, A. (1972). *The Call Girls.* Hutchinson, London.

Lane, M. (1978). *The Magic Years of Beatrix Potter.* Frederick Warne, London.

Lightman, A. (1992). *Einstein's Dreams.* Bloomsbury, London.

McCrum, R., Cran, W. and MacNeil, R. (1986). *The Story of English.* Faber and Faber, London.

Marhall, D. L. and Folsom, M. W. (1991). 'Mate choice in plants: an anatomical to population perspective', *Annual Review of Ecology and Systematics* 22: 37–64.

Mulvey, J. (1979). 'The new frontier of particle physics', *Nature, London* 278: 403–409.

Myers, G. (1985). 'Texts as knowledge claims: the social construction of two biologists' articles', *Social Studies of Science* 15: 593–630.

Myers, G. (1990). *Writing Biology.* University of Wisconsin Press, Madison, WI.

RACTER (1985). *The Policeman's Beard is Half Constructed.* Warner Books, New York.

Schwartz, J. (1992). *The Creative Moment: How Science Made Itself Alien to Modern Culture.* Cape, London.

Small, M. (1990). [Review of D. Haraway, *Primate Visions*], *American Journal of Physical Anthropology* 82: 527–532.

Wilson, E. O. (1975). *Sociobiology: The New Synthesis.* Belknap Press, Cambridge, MA.

Chapter 9

Bowler, P. J. (1986). *The Idea of Evolution.* University of California Press, Los Angeles.

Carleton, D. (1992). 'The possibility of mastery', *Nonesuch* (Spring): 21–31 (University of Bristol).

Collier, A. (1989). *Scientific Realism and Socialist Thought.* Harvester Wheatsheaf: Hemel Hemstead.

Coplestone, F. (1963). *A History of Philosophy,* vol. 7: *Fichte to Nietzsche.* Burns Oates, London.

Dennett, D. (1991). *Consciousness Explained.* Allen Lane, London.

Gale, G. (1979). *Theory of Science.* McGraw-Hill, Toronto.

Gasman, D. (1971). *The Scientific Origins of National Socialism.* Elsevier, New York.

Gellner, E. (1992). *Postmodernism, Reason and Religion.* Routledge, London.

Gross, P. R. and Levitt, N. (1993). *Higher Superstition: The Academic Left and its Quarrel with Science*. Johns Hopkins University Press.

Haraway, D. (1989). *Primate Visions*. Routledge, New York.

Hesse, M. (1980). *Revolutions and Reconstructions in the Philosophy of Science*. Harvester, Brighton.

Horton, R. (1993). *Patterns of Thought in Africa and the West: Essays on Magic, Religion and Science*. Cambridge University Press, Cambridge.

Huff, T. (1993). *The Rise of Early Modern Science*. Cambridge University Press, Cambridge.

Huxley, A. (1952). *The Devils of Loudun*. Chatto & Windus, London.

Leach, E. (1954). *Political Systems of Highland Burma*. London.

Lepenies, W. (1989). 'The direction of the disciplines: the future of the universities', *Comparative Criticism* 11: 00–00.

Lynch, M. (1982). 'Technical works and critical enquiry: investigations in a scientific laboratory', *Social Studies in Science* 12: 499–533.

Mayr, E. (1982). *The Growth of Biological Thought*. Belknap Press, Cambridge, MA.

Merton, R. (1973). *The Sociology of Science*. University of Chicago Press, Chicago.

Mulkay, M. (1979). *Science and the Sociology of Knowledge*. Allen & Unwin, London.

Myers, G. (1986). *Writing Biology*. University of Wisconsin Press, Madison, WI.

Popper, K. R. (1962). *The Open Society and its Enemies*. Routledge & Kegan Paul, London.

Pringle, S. (1992). 'A scientific dinner party', *Nonesuch* (Spring), p. 47 (University of Bristol).

Shapin, S. (1982). 'History of science and its sociological reconstructions', *History of Science* 20: 157–211.

Shapin, S. and Schaffer, S. (1984). *Leviathan and the Airpump: Hobbes, Boyle, and the Experimental Life*. Princeton University Press, Princeton, NJ.

Woolgar, S. (1980). 'Discovery: logic and sequence in a scientific text', *The Social Process of Scientific Investigation*, ed. K. Knorr, R. Krohn and R. Whitley, vol. 4: 239–268. Reidel, Dortrecht.

Zmarzlik, H.-G. (1972). 'Social Darwinism in Germany seen as historical problem', *Republic and Reich: The Making of the Nazi Revolution*, ed. H. Holborn: 435–474. Pantheon, New York.

Chapter 10

Boyd, R. and Richerson, P. (1986). *Evolution and the Cultural Process*. Chicago University Press, Chicago.

Cromer, A. (1993). *Uncommon Sense*. Oxford University Press, Oxford.

Hughes, P. (1993). 'All part of life's rich tapestry', *New Scientist* (4 Dec): 47–48.
Kaye, H. (1986). *The Social Meaning of Modern Biology*. Yale University Press, New Haven, CT.
McGovern, T. H. (1981). 'The economics of extinction in Norse Greenland', *Climate and History*, ed. T. M. L. Wrigley, M. J. Ingram and C. Farmer: 404–433. Cambridge University Press, Cambridge.
National Science Foundation (1994). *Science and Engineering Indicators—1993*. National Science Foundation, Washington, D. C.
Rose, S., Kamin, L. J. and Lewontin, R. C. (1984). *Not in Our Genes: Biology, Ideology and Human Nature*. Penguin, Harmondsworth.
Segestrale, U. (1986). 'Colleagues in conflict: an "in vivo" analysis of the sociobiology controversy', *Biology and Philosophy* 1: 53–87.

Index